AF357827

Computational Approaches for Protein Dynamics and Function

Computational Approaches for Protein Dynamics and Function

Editors

Robert Jernigan
Domenico Scaramozzino

Basel • Beijing • Wuhan • Barcelona • Belgrade • Novi Sad • Cluj • Manchester

Editors

Robert Jernigan
Iowa State University
Ames
USA

Domenico Scaramozzino
Karolinska Institutet
Solna
Sweden

Editorial Office
MDPI AG
Grosspeteranlage 5
4052 Basel, Switzerland

This is a reprint of articles from the Special Issue published online in the open access journal *Applied Sciences* (ISSN 2076-3417) (available at: https://www.mdpi.com/journal/applsci/special_issues/Proteins_Dynamics).

For citation purposes, cite each article independently as indicated on the article page online and as indicated below:

Lastname, A.A.; Lastname, B.B. Article Title. *Journal Name* **Year**, *Volume Number*, Page Range.

ISBN 978-3-7258-2555-4 (Hbk)
ISBN 978-3-7258-2556-1 (PDF)
doi.org/10.3390/books978-3-7258-2556-1

Contents

About the Editors

Robert Jernigan

Prof. Dr. Robert Jernigan was an undergraduate at CalTech in Chemistry and received his Ph.D. from Stanford University, where he worked on computational polymer physics with the Nobel Laureate Paul Flory. Following a postdoc at UCSD and a short postdoc at NIH, he was recruited into a new tenured independent investigator position at the National Cancer Institute in Bethesda. There, he was Section Head of the Molecular Structure Section and Deputy Chief of the Laboratory of Experimental and Computational Biology for many years, with outstanding research support. In 2002, he left to become a Professor at Iowa State University in the Department of Biochemistry, Biophysics and Molecular Biology, as well as the Director of the endowed Laurence H. Baker Center for Bioinformatics and Biological Statistics, where he is now a Curtiss Distinguished Professor. He has applied physical principles at a basic level to a wide range of molecular problems related to biomedicine. These have focused on basic methods that are able to treat proteins and nucleic acids, which usually involve coarse-graining the structures so that structures and sequences can be directly integrated. His collective work demonstrates that atoms are not always needed to comprehend many of the behaviors and mechanisms of proteins. This point of view is critical to facilitating the progression of simulations for the ever-larger structures being uncovered. His simulations of the ribosome uncovered mechanisms that have the remarkable ability to provide new insights. He believes strongly that simulations in biology provide a computational platform for the development of new hypotheses. Some of his work provides important foundations for protein design. His outstanding accomplishments in research have had a highly significant and important impact on the field of computational molecular biology.

Domenico Scaramozzino

Dr. Domenico Scaramozzino pursued his education in Structural Engineering at Politecnico di Torino (Italy), where he was awarded his PhD in Civil and Environmental Engineering in 2021. After conducting research on tall buildings and civil lattice structures, he became fascinated by a completely different type of structure that is not made of steel and concrete but of backbone atoms and side chains: proteins. During his PhD, he visited the *Jernigan Lab* at Iowa State University (US) in order to conduct research on protein dynamics. Currently, he works as a postdoctoral scholar in Computational Structural Biology in the *Protein Dynamics and Mutation Lab* at Karolinska Institutet (Sweden). He is interested in understanding how proteins achieve biological functionality by exploiting their structure, flexibility, and conformational dynamics. In order to achieve this purpose, he is developing new computational methods using elastic network modeling, normal mode analysis, and coarse-grained sampling algorithms. He is the lead author of the book *Waves in Biomechanics: THz vibrations and modal analysis in proteins and macromolecular structures*, and he was recently awarded the *ICCSE Best Young Researcher Award*.

Preface

In recent years, we have witnessed two major revolutions in the field of protein structural biology. On one hand, experimental techniques are enhancing the ease with which the structural details of complex protein systems are visualized with an astonishingly high resolution. More than 200,000 experimental structures are currently deposited in the Protein Data Bank (PDB), providing us with a vast volume of data with which to decipher the protein structure–function relationship. Cryogenic electron microscopy (cryoEM) has offered significant advances in this field, enabling increasingly large proteins to be resolved at increasingly high resolutions. On the other hand, significant advancements in machine learning (ML) and artificial intelligence (AI) have allowed huge amounts of sequence and structural data to be mined, analyzed, and interpreted in a shorter time. AI algorithms, such as AlphaFold, have been particularly valuable for the prediction of protein structures, providing us with more than 1,000,000 new computer-generated structures in the PDB. Yet, despite this huge volume of experimental and computational data, several fundamental questions remain unanswered. How do proteins exploit their structural and dynamical features to initiate biological activity? What triggers the conformational changes required for protein function? Does the external environment play a role in driving protein-relevant dynamics? How do signals associated with protein–ligand and protein–protein binding propagate to generate allosteric mechanisms? Do deleterious mutations disrupt protein functionality by impairing the biologically relevant motions?

In this Special Issue, we have compiled recent works that attempt to answer these (and other) relevant questions. All included research studies make use of one or more computational techniques in order to elucidate the intimate relationship between the sequence, structure, dynamics, and function of proteins. The reader will find that these techniques can range from molecular dynamics (MD) to normal mode analysis (NMA), elastic network models (ENMs), the protein contact network (PCN), and many more. We believe that the content of this Special Issue is particularly beneficial for the computational and structural biology communities; it creates knowledge on the dynamics and function of proteins and hopes to stimulate new and exciting research questions.

Robert Jernigan and Domenico Scaramozzino

Editors

Editorial

Special Issue on "Computational Approaches for Protein Dynamics and Function"

Domenico Scaramozzino [1,*] and Robert L. Jernigan [2]

[1] Department of Oncology-Pathology, Karolinska Institutet, 171 65 Solna, Sweden
[2] Department of Biochemistry, Biophysics and Molecular Biology, Iowa State University, Ames, IA 50011, USA; jernigan@iastate.edu
* Correspondence: domenico.scaramozzino@ki.se

Citation: Scaramozzino, D.; Jernigan, R.L. Special Issue on "Computational Approaches for Protein Dynamics and Function". *Appl. Sci.* **2023**, *13*, 8522. https://doi.org/10.3390/app13148522

Received: 10 July 2023
Accepted: 20 July 2023
Published: 24 July 2023

Proteins are fundamental macromolecules that sustain living organisms by performing an astonishingly wide variety of tasks. They adopt extremely diverse shapes that perform highly specific functions, achieved through temporal optimization over millions of years of evolution. Proteins usually have specific flexibility that enables them to undergo the conformational changes necessary to perform certain functions. Understanding protein flexibility and conformational dynamics is thus pivotal to determine how proteins work. In the era of advanced computing technologies, can we use computational approaches to elucidate how proteins' structures and dynamics drive their function? This Special Issue collects recent studies that employ different computational methods to answer this question.

Molecular Dynamics (MD) is regarded as the gold standard when it comes to protein dynamics. Molecular docking and virtual screening are popular computational methods to discover novel drugs and protein inhibitors. Singh et al. [1] combined docking-based virtual screening with MD simulations to find potential inhibitors of Mycobacterium tuberculosis Fatty Acid Synthase type-I (Mtb FAS-I). By screening a database of ~55,000 compounds, the authors narrowed their targets down to nine potential candidates. By carrying out short MD simulations and binding energy calculations for the nine protein–ligand complexes, the authors reduced their targets to four molecules that might act as pioneer FAS-I inhibitors, paving the way to a novel treatment for tuberculosis.

Simpler than MD simulations, Elastic Network Models (ENMs) simulate protein dynamics and flexibility by modeling the protein as a network of elastic springs, and are often used in combination with Normal Mode Analysis (NMA). Scaramozzino et al. [2] introduced a dynamic solvent effect into ENMs to more effectively reproduce X-ray fluctuations than using solvent-free ENMs. By investigating a dataset of ~1k protein structures, they showed that the highest correlation with experimental data was obtained when random perturbations were applied to the solvent-exposed surface and when water molecules were included into the ENM. These findings suggest that a tightly bound water layer is important for modulating protein flexibility, and that protein fluctuations likely originate during the bombardment of the structure by the solvent.

ENMs were also used by Tarenzi et al. [3] to decipher structure–dynamics–function relationships. ENM-NMA was applied to a dataset of 116 different proteases, and proteins were clustered together based on their "dynamic distance" in the space of normal modes. Proteins that belonged to the same sub-families, and thus, had similar sequences and functions, also had similar dynamics. Interestingly, some sub-families were also clustered together in certain cases, suggesting that they might share similar dynamic traits despite having different evolutionary origins. This method also built a basis of dynamic vectors that could describe the most important features of the large-scale motions in the dataset and was validated by MD.

Structural modeling, protein–protein docking, and Protein Contact Networks (PCNs) were used by Drago et al. [4] to analyze the interactions between the von Willebrand

Factor (VWF) and Factor VIII belonging to the coagulation cascade. Two models of FVIII (full-length and without the B-domain) were docked with VWF. The binding energies and PCN results were subsequently analyzed to assess the stability of the FVIII-VWF interfaces and find potential allosteric pathways. The results showed that the A3-C1 domains are the preferential binding sites for VWF. This agrees with the experimental structure of efanesoctocog alfa, a novel (B-domain free) FVIII-VWF complex used in medication for hemophilia A.

The effect of ligand binding on the dynamics and allosteric pathways in human Glutathione Transferase A1 (GTSA1) was investigated by Nicolaï et al. [5]. MD simulations were carried out on apo GTSA1, and on GTSA1 bound to glutathione (GSH) or to a GS-conjugate ligand. Free-energy surfaces and 1D profiles were reconstructed based on the variability of two sets of coarse-grained angles. By looking at the differences between free-energy landscapes, the authors recognized 11 residues known to be key in ligand binding and identified 22 more that were previously unknown. Some of these residues are distant from the binding sites, highlighting the importance of long-range allosteric effects for protein-ligand interactions.

MD simulations were also used by Sogunmez and Akten [6] to analyze the dynamics of human β_2-adrenergic receptor (β_2AR) in complex with a G-protein, and its signal transmission in its fully active state. Mutual information and transfer entropy were used to infer correlations between C^α displacements and the rotameric states of the backbone and side-chain angles. The use of rotameric states enabled the recognition of strong correlations in almost all loop regions; the authors identified the loops as potential allosteric hot spots and highlighted the donor nature of polar residues and their importance in signal transmission.

The intertwined relationship between protein sequence, structure, dynamics, and function was broadly addressed by Orellana in a perspective article [7]. After a review of the literature, the author highlighted specific examples in which functional motions are conserved from bacteria to mammals. Emblematic of this is the mammalian proton exchanger NHE9, which shares only 20% sequence similarity with its distant bacterial homologs but exhibits a remarkably high overlap (~70–90%) in terms of functional motions, as assessed via Principal Component Analysis (PCA) of experimental ensembles and NMA. This is evidence that protein motions are a key phenotype selected during evolution. The author argues that cancer might also adopt this strategy to favor mutations that disrupt functional motions, supporting the emerging notion that disease mutations often affect protein dynamics.

Author Contributions: Conceptualization: D.S. and R.L.J.; Curation of the Special Issue: D.S. and R.L.J.; Writing of the Editorial—original draft preparation: D.S.; Writing of the Editorial—review and editing: D.S. and R.L.J. All authors have read and agreed to the published version of the manuscript.

Acknowledgments: We would like to thank all the authors and reviewers for their valuable contributions to the Special Issue 'Computational Approaches for Protein Dynamics and Function'.

Conflicts of Interest: The authors declare no conflict of interest.

References

1. Singh, N.; Mao, S.-Q.; Li, W. Identification of Novel Inhibitors of Type-I Mycobacterium Tuberculosis Fatty Acid Synthase Using Docking-Based Virtual Screening and Molecular Dynamics. *Appl. Sci.* **2021**, *11*, 6977. [CrossRef]
2. Scaramozzino, D.; Khade, P.M.; Jernigan, R.L. Protein Fluctuations in Response to Random External Forces. *Appl. Sci.* **2022**, *12*, 2344. [CrossRef]
3. Tarenzi, T.; Mattiotti, G.; Rigoli, M.; Potestio, R. In Search of a Dynamical Vocabulary: A Pipeline to Construct a Basis of Shared Traits in Large-Scale Motions of Proteins. *Appl. Sci.* **2022**, *12*, 7157. [CrossRef]
4. Drago, V.; Di Paola, L.; Lesieur, C.; Bernardini, R.; Bucolo, C.; Platania, C.B.M. In-Silico Characterization of von Willebrand Factor Bound to FVIII. *Appl. Sci.* **2022**, *12*, 7855. [CrossRef]
5. Nicolaï, A.; Petiot, N.; Grassein, P.; Delarue, P.; Neiers, F.; Senet, P. Free-Energy Landscape Analysis of Protein-Ligand Binding: The Case of Human Glutathione Transferase A1. *Appl. Sci.* **2022**, *12*, 8196. [CrossRef]

6.   Sogunmez, N.; Akten, E.D. Information Transfer in Active States of Human β_2-Adrenergic Receptor via Inter-Rotameric Motions of Loop Regions. *Appl. Sci.* **2022**, *12*, 8530. [CrossRef]

7.   Orellana, L. Are Protein Shape-Encoded Lowest-Frequency Motions a Key Phenotype Selected by Evolution? *Appl. Sci.* **2023**, *13*, 6756. [CrossRef]

Article

Identification of Novel Inhibitors of Type-I Mycobacterium Tuberculosis Fatty Acid Synthase Using Docking-Based Virtual Screening and Molecular Dynamics Simulation

Nidhi Singh [1,2], Shi-Qing Mao [1] and Wenjin Li [1,*]

[1] Institute for Advanced Study, Shenzhen University, Shenzhen 518060, China; tanwar.nidhi7@gmail.com (N.S.); msq@szu.edu.cn (S.-Q.M.)
[2] College of Physics and Optoelectronic Engineering, Shenzhen University, Shenzhen 518060, China
[*] Correspondence: liwenjin@szu.edu.cn; Tel.: +86-755-26942336

Abstract: Mycobacterial fatty acid synthase type-I (FAS-I) has an important role in the de novo synthesis of fatty acids, which constitute a major component of the cell wall. The essentiality of FAS-I in the survival and growth of mycobacterium makes it an attractive drug target. However, targeted inhibitors against Mycobacterial FAS-I have not been reported yet. Recently, the structure of FAS-I from Mycobacterium tuberculosis was solved. Therefore, in a quest to find potential inhibitors against FAS-I, molecular docking-based virtual screening and molecular dynamics simulation were done. Subsequently, molecular dynamic simulations based on binding free energy calculations were done to gain insight into the predicted binding mode of putative hits. The detailed analysis resulted in the selection of four putative inhibitors. For compounds BTB14738, RH00608, SPB02705, and CD01000, binding free energy was calculated as -72.27 ± 12.63, -68.06 ± 11.80, -63.57 ± 12.22, and -51.28 ± 13.74 KJ/mol, respectively. These compounds are proposed to be promising pioneer hits.

Keywords: fatty acid synthase; molecular docking; virtual screening; molecular dynamics simulations; MM/PBSA; binding free energy

Citation: Singh, N.; Mao, S.-Q.; Li, W. Identification of Novel Inhibitors of Type-I Mycobacterium Tuberculosis Fatty Acid Synthase Using Docking-Based Virtual Screening and Molecular Dynamics Simulation. *Appl. Sci.* **2021**, *11*, 6977. https://doi.org/10.3390/app11156977

Academic Editor: Robert Jernigan

Received: 30 June 2021
Accepted: 27 July 2021
Published: 29 July 2021

Publisher's Note: MDPI stays neutral with regard to jurisdictional claims in published maps and institutional affiliations.

1. Introduction

Tuberculosis (TB) is a leading health problem worldwide. According to the World Health Organization estimation, 10 million new cases were reported in 2018 alone, and 1.5 million people have died of it [1]. Tuberculosis is contagious and an airborne disease caused by Mycobacterium tuberculosis. The current drug regimen for the treatment of TB relies upon a six-month course of anti-microbial drugs [2]. The lengthy regimen leads to non-adherence and consequently the emergence and spread of drug-resistant strains. The rise of multi drug-resistant strains and co-occurrence with HIV also pose challenges in combating mycobacterium [2].

Mycobacterium has successfully evaded the host system since ancient times. Insight into the success story of Mycobacterium shows that virulence is largely attributed to its thick layer of mycolic acids, a major component of the cell wall [3–5]. It acts as an efficient barrier due to low permeability and fluidity and provides intrinsic resistance to anti-microbial drugs. The lipid biosynthesis in Mycobacterium is carried out by a combination of two enzymatic systems—FAS-I and FAS-II. Mycolic acids are long fatty acids and characterized by hydrophobic C54-C63 fatty acids with C22-C24 side chains in Mycobacterium [6]. The FAS-II system is comprised of four discrete enzymes, which work successively and repetitively to elongate the acyl chain, similar to the system found in prokaryotes and plants [7]. On the other hand, FAS-I is a multi-domain and multi-functional enzyme similar to fungi and higher eukaryotes [6,8]. It catalyses the de novo synthesis of fatty acids starting from acetyl-CoA and is capable of elongating fatty acids up to C24/26 [9]. The fatty acid chain is further elongated to meromycolate (C56) through the FAS-II system and later condensed with C26, resulting in the formation of mycolic acids [6,7]. Furthermore, fatty acid synthesis has been reported to be essential in mycobacteria [10,11]. The importance of fatty acid synthesis is also manifested by the use of drugs isoniazid, ethambutol, and pyrazinamide in the current drug regimen, which are inhibitors of mycolic acid biosynthesis [12]. Isoniazid and ethambutol targets enoyl reductase domain of the FAS-II system to inhibit mycolic acid synthesis and be used as first- and second-line drugs against TB. Pyrazinamide is being used as a first line drug and has a key role in shortening the drug regimen from nine to six months [13]. Moreover, analogs of pyrazinamide have been reported to target the FAS-I system of Mycobacterium [14,15]. The role of mycolic acids in forming the cell wall and its key role in growth and survival of mycobacterium makes the FAS system an attractive drug target. Recently the structure of mycobacterial FAS-I was elucidated by Nadav Elad et al. [16]. This has paved the way for structure based inhibitor identification against the mycobacterial FAS-I system.

The receptor-inhibitor design is the spirit of any drug design process and the information of receptor-ligand complex can be channelized through many ways. Amongst them, virtual screening is one of the most commonly used to discover novel scaffolds and lead compounds [17–19]. Molecular docking is a popular choice to carry out the virtual screening and has proved its mettle in hit identification and lead optimization. Molecular docking has been successfully implemented to screen large compound libraries against the drug targets and identification of mechanism of action of known active compounds [20–22]. Docking methods have been used to screen in-house as well as commercial libraries. For example, virtual screening was successfully employed for the identification of antibacterial inhibitors against NAD synthetase [23]. Docking-based virtual screening was done to identify novel inhibitors against leishmanial nucleoside diphosphate kinases [24]. In a recent study, docking-based screening was performed to identify potential inhibitors against isocitrate lyase of Mtb [25]. There are several other studies in which docking-based virtual screening has been successfully applied for the identification of novel inhibitors [26–30]. The Molecular Mechanics Poisson-Boltzman Surface Area (MM/PBSA) method is used to estimate binding affinity of protein-ligand complexes predicted by the molecular docking. The successful applications of MM/PBSA in virtual screening protocols has been reviewed by Giulio Poli et. al. [31].

To date, inhibitors of enoyl reductase domain of Mtb FAS-I have not been reported. Owing to the crucial role of FAS-I and available structural information, we were intrigued to search potential inhibitors against mycobacterial FAS-I. For this, molecular docking-based virtual screening protocol and molecular dynamics-based MM/PBSA calculations were implemented to identify putative hit compounds. The binding mode of active compounds has been proposed through molecular docking. The proposed inhibitors are pioneers and can serve as the basis for the design and optimization of new inhibitory compounds against TB.

2. Materials and Methods

2.1. Ligand Library Preparation

The commercial maybridge screening library [32] was selected to perform virtual screening. The selected maybridge collection consist of compounds which obey Lipinski's "rule of five"; hence, demonstrating good ADME parameters. In addition, screening collection represents over 87% pharmacophores in the world drug index. Therefore, the hits obtained can undergo further development. The screening library is available online and was downloaded from the website in sdf format. Firstly, the screening library was prepared for docking using the "surflex for searching" protocol of Sybyl 2.1 software. It follows the general clean-up and one least strained energy 3D conformer generation steps. The compounds collection obtained (54,646) was saved and used for docking-based virtual screening.

2.2. Docking-Based Screening

Surflex Docking: The Mtb FAS-I structure in complex with FMN was retrieved from the PDB database (PDBID: 6GJC). FAS-I is a large α6 subtype complex consisting of six long polypeptide chains. Each chain is 3069 amino acids long and contains seven catalytic domains. For primary screening and docking, the Surflex-Dock program of sybyl 2.1 was used in screen mode [33]. For the receptor preparation step, only chain A was retained, while the other chains and water molecules were removed. The FMN molecule was also retained as it is found tightly bound to FAS-I. In addition, hydrogen atoms were added, and atom types and AMBER charges were assigned to protein atoms. Ligand NADPH was extracted from the structure of FAS-I of thermomyces lanuginosus (PDB:4V59), which is homologous to mycobacterial FAS-I. Ligand-based protomol was generated using NADPH, keeping the threshold 0.5 and bloat set to default. The molecular docking was done with default settings. In the next step, re-ranking of high scoring hits was done using the Geom-X mode. The spin density for search is higher in Geom-X mode and set to the value of nine, while in screen mode it is three. Therefore, accuracy in ranking the hits on the basis of docking score is enhanced as the search becomes denser. Surflex is based on the Hammerhead procedure for docking the flexible ligands into the binding site of the receptor. It is based upon the generation of ligand fragments and their alignment onto the identified probes; the remaining fragments of ligands were then docked. The scoring function is empirical and derived through a weighted sum of non-linear functions of protein-ligand atomic Van der Waals surface distances. Hydrophobic, entropic, polar, solvation, repulsive, and crash terms are included in the scoring function. The score predicts binding affinity in −log 10(Kd) units.

2.3. Molecular Dynamics Simulation

To assess the suitability of the selected hits after docking score filter in terms of stability of protein-ligand interactions, molecular dynamics simulations were carried out using GROMACS 5.0 software [34]. The topology for ligands was generated through the cGenff server [35,36]. The CHARMM36 force field was applied to the protein-ligand systems. The system was kept in a cubic box and placed 10 Å from the box edge. The TIP3P water model was used for protein-ligand complexes. The whole system was neutralized by adding appropriate ions and energy minimization was done using the steepest descent algorithm. In the next step, energy minimized systems were subjected to NVT and NPT equilibration phases for 100 ps each. Isotropic pressure coupling was performed using the Parrinello-Rahman method, keeping the pressure coupling time at 2 ps and isothermal compressibility 4.5×10^{-5} bar^{-1}. Electrostatic interactions were treated with Particle Mesh Ewald method [37]. Coulomb and Van der Waals interactions were truncated at 1 nm. The systems were subjected to production run for ten ns each and at every ten ps, conformations were saved. The gromacs rms utilities were employed to calculate root mean square deviation (r.m.s.d) of ligands. The GRACE program was used to plot the graphs [38]. The molecular dynamics simulation was used to calculate protein-ligand interaction energy based upon the MM/PBSA method.

2.4. MM/PBSA-Based Protein–Ligand Interaction Energy Calculation

In order to calculate the binding energy for predicted complexes, the MM/PBSA method was applied. The simulation-based end point methods, such as molecular mechanics with Poisson–Boltzmann (MM/PBSA) and molecular mechanics with generalized Born and surface area (MM-GBSA) due to their computational efficiency [39–42]. For all complexes, binding free energy calculations was carried out using the g_mmpbsa tool [40]. The binding free energy is given by:

$$\Delta G_{bind} = \langle \Delta E_{MM} \rangle - T \Delta S + \langle \Delta G_{solvation} \rangle \tag{1}$$

$$\Delta E_{MM} = \Delta E_{bonded} + \Delta E_{vdW} + \Delta E_{elec} \tag{2}$$

$$\Delta G_{solvation} = \Delta G_{polar} + \Delta G_{nonpolar} \tag{3}$$

Here, ΔE_{MM} represents the energy of bonded and non-bonded terms and is calculated on the basis of molecular mechanics force-field parameters. In addition, in the single trajectory approach, the protein ligand conformation in bound and unbound form is identical and therefore is assumed to be zero [43]. The solvation free energy term includes polar and non-polar terms. The polar solvation energy is solved using the Poisson-Boltzmann equation [44] while non-polar solvation energy is calculated by attractive and repulsive forces between the solute and solvent, generated through cavity formation and Van der Waals interactions [45,46]. For the current study, MM/PBSA calculations were done on the last 5 ns segment of the trajectory. For each system, 100 snapshots were extracted at the interval of 50 ps along the trajectory.

3. Results

3.1. Structural Model of NADPH-Bound Mtb FAS-I

Mtb FAS-I is a large α6 subtype barrel-shaped complex consisting of six long polypeptide chains (α chains). Each α chain is 3069 amino acids long and comprised of seven catalytic domains [16]. The overall architecture depicts the whole complex as a central wheel capped by domes on each side. The seven catalytic domains are namely acetyltransferase, enoyl reductase (ER), dehydratase, malonyl transacylase, ketoacyl reductase, ketoacyl synthase, and acyl carrier protein. The Mtb FAS-I was found to be similar to fungal FAS which is a homolog that retains the barrel shape complex [37]. The comparison of catalytic clefts between mycobacterial and fungal domains is given in detail [16]. Furthermore, the crystal structure of fungal FAS-I has been reported in complex with NADP+ and FMN (PDB:4V59). The structural comparison revealed that the enoyl reductase domain of Mtb has a wider catalytic cleft and FMN is more exposed due to local amino acid composition. The ER domain is involved in the catalysis of the last step of fatty acid elongation cycle through FMN-dependent reduction of enoyl-ACP intermediate to saturated acyl-ACP. The ER is embedded in the FAS-I complex so that it allows easy access of NADPH to the binding site from outside of the FAS-I complex, while the catalytic centre is accessible from the inside of the reaction chamber. A two-step ping-pong mechanism had been proposed for catalytic mechanism of the ER of fungal FAS-I [47]. Moreover, analogs of pyrazinamide have been reported to be competitive inhibitors of NADPH binding to Mtb FAS-I [14]. Hence, we were intrigued to identify putative inhibitors of Mtb FAS-I. The availability of the Mtb FAS-I structure paved the way for docking-based screening. Taking in consideration the close homology with fungal FAS-I, NADPH was extracted from fungal FAS-I and docked to the binding site of Mtb FAS-I (Figure 1a). The residues involved in the binding site of NADPH were studied and compared (Table 1) to the NADPH binding site in fungal FAS-I. The residues were found to be conserved and thus indicates the suitability of NADPH binding. The residues Tyr636, Lys1026, and pro1027 hold adenosine moeity; Asp 1045 interacts with pyrophophate part, while His 751 acts as a catalytic residue in fungal FAS-I. Similarly, the corresponding residues in Mtb FAS-I have been tabulated (Table 1). Notably, the residue His584 in Mtb seems to act as a catalytic residue and can play an important role in the function of the enoyl reductase domain.

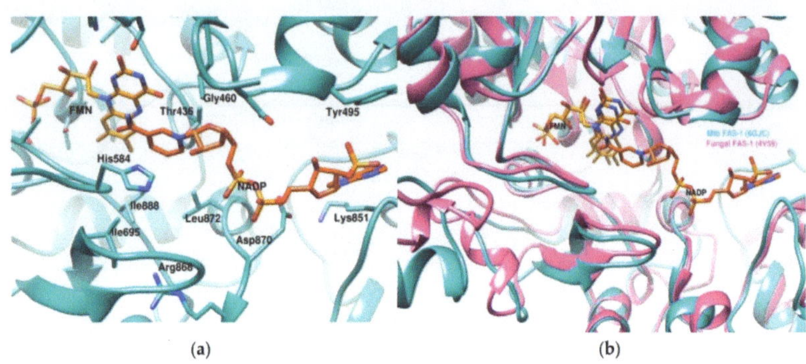

Figure 1. (**a**) Mtb FAS-I shown in cyan color with bound FMN and docked NADP (**b**) The enoyl reductase domain of Mtb FAS-I (cyan color, PDB:6GJC) superimposed onto Thermomyces lanuginosus FAS-I (pink color, PDB:4V59).

Table 1. The residue comparison for the NADPH binding site of enoyl reductase between fungal and Mtb FAS-I.

NADPH Binding Site Residues from Thermomyces Lanuginosus (PDB Code: 4V59)	NADPH Binding Site Residues from Mtb (PDB Code: 6GJC) Obtained by Docking
Tyr636	Tyr495
Asp668	-
Asp952	-
Lys1026	Lys851
Pro1027	Pro852
Asp1045	Asp870
Ser1046	Ser871
Lys1044	Ser869
Leu1047	Leu872
Thr609	Thr436
Ile663	Leu490
Gly749	Gly582
Gly750	Gly583
His751	His584
Glu863	-

3.2. Docking-Based Screening

To discover novel Mtb FAS-I inhibitors, the cryo-EM structure of FAS-I in complex with FMN (PDB ID: 6GJC) was used for molecular docking-based virtual screening utilizing the Surflex-dock module of the Sybyl 2.1 software. The NADP binding pocket was used for performing docking-based screening, comprised of residues Met435, Thr436, Pro437, ValL440, Ala458, Gly460, GLY583, His584, His585, Ala693, Asp694, Ile695, Pro852, Arg868, Ser869, Asp870, Ser871, Leu872, Trp873, Gln874, and the FMN molecule. To address accuracy and efficiency, we carried out the screening protocol in a hierarchical strategy summarized in a work-flow diagram (Figure 2).

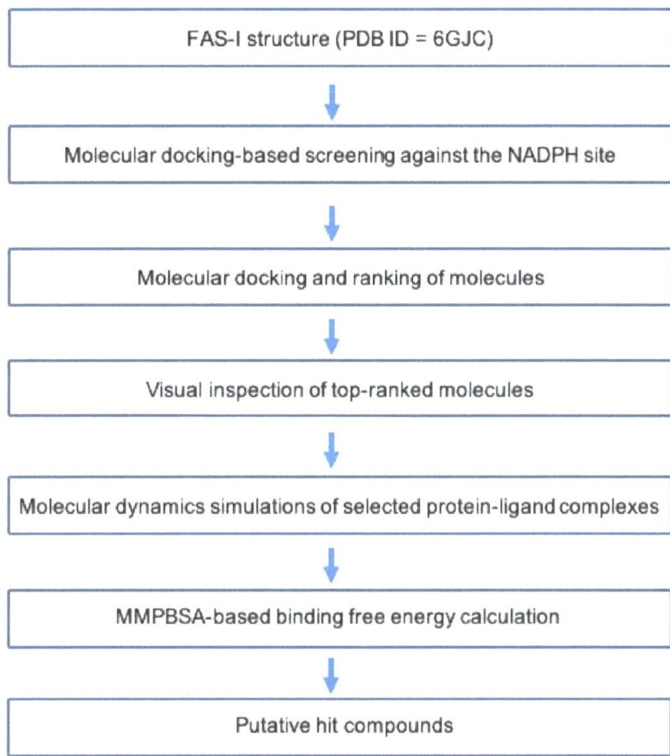

Figure 2. The work plan for molecular docking-based virtual screening.

Primary screening: The Maybridge screening library consisting of 54,646 molecules was docked using Surflex-dock into the active site of Mtb FAS-I. The top scoring molecules with a score equal to or higher than the score (8.0) for the binding site were considered for the next step. The choice of cut-off value in this study was guided by reproduction of docking pose of phenylimidazole derivative inhibitor enoyl–ACP reductase (FabK) from Streptococcus pneumoniae, which is competitive inhibitor of NADPH (PDB:2Z6J). This cut-off resulted in 528 molecules, which were docked again using Surflex-GeomX mode for re-ranking and to improve pose accuracy. Subsequently, the top 150 molecules were inspected visually for favorable interactions.

Visual inspection: The hits from GeomX were inspected visually for their binding mode(s) for further selection. The following criteria were considered: (1) π-π stacking interaction between the ligand and the FMN molecule; (2) the interaction with residues His584 and Thr436; (3) the formation of hydrogen bonds and other hydrophobic interactions; and (4) the stability in docked pose and fitness of molecule in the binding site. This step was primarily used to enhance specificity and eliminate the compounds having higher score due to interactions with other residues. Based on these criteria, a total of nine molecules were selected for subsequent molecular dynamics simulations. The structures of these compounds are shown in Figure 3.

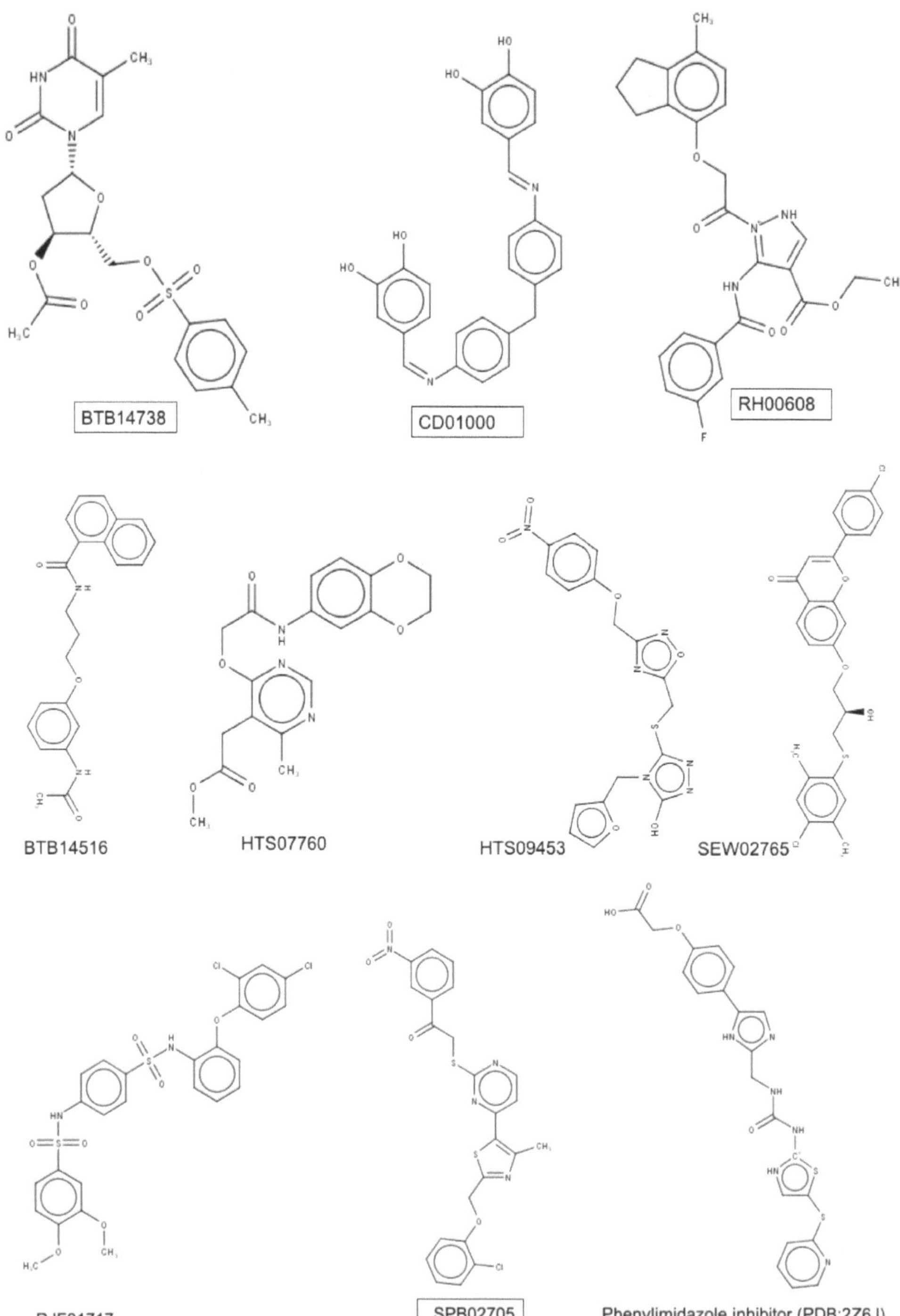

Figure 3. The structures of nine selected compounds after docking-based screening are shown along with the inhibitor of *S. pneumoniae* Enoyl-Acyl Carrier Protein Reductase (FabK). The labels of four putative inhibitors are highlighted in boxes.

3.3. Molecular Docking and Proposed Mode of Binding of Putative Hits

The predicted binding mode of compounds selected on the basis of molecular docking and visual inspection are discussed in this section. The detailed 2D interaction plots for all the complexes are given in detail (Figure S1), while the 3D interaction plots for protein-ligand complexes are shown in Figure 4. The compound BTB14738 showed hydrogen bonding with residues Thr436 and Arg868 and strong π-π stacking with FMN molecule. The residues Leu872 and Ile695 are involved in hydrophobic interaction, while the sulfur atom can be involved in π-sulfur interaction with residue His584 (Figure 4a). The compound BTB14516 showed hydrogen bonds with the residues Thr436, Gly460, and His584 and π-π stacking with the FMN molecule (Figure 4b). The compound CD01000 showed hydrogen bonds with the residues Arg868, Gln690, and Lys550 and π-π stacking with the FMN molecule (Figure 4c). The compound SEW02765 showed extensive interactions in the form of hydrogen bonds with the residues Thr436 and Arg868 and π-sulfur with His584. The residues Leu490, Ala524, Ile888, Ile695, and Leu872 provided hydrophobic contacts (Figure 4d). The compound HTS07760 showed favorable interactions in the form of hydrogen bonds with the residues Thr436, His584, and Arg868 and π-π stacking with the FMN molecule (Figure 4e). The compound HTS09453 showed hydrogen bonds with residues His584, Thr436, Arg868, and Ser871 and π-π stacking with the FMN molecule (Figure 4f). The next compound RH00608 showed hydrogen bond with the residues Thr436, His584, Arg868, and halogen bond with the residue Asp870 and π-π interaction with FMN. It is stabilized through various hydrophobic contacts to Ala524, Ala581, Ile695, Ile888, and Leu872 (Figure 4g). The compound RJF01717 showed extensive hydrogen bonds with residues His584, Thr436, Ser523, Gln690, Ser871 and Arg868 and retaining π-π stacking with FMN (Figure 4h). The compound SPB02705 showed π-π stacking with residue His584 and the FMN molecule through two aromatic rings and hydrogen bonds with residues His584 and Arg868 (Figure 4i). In brief, the compounds showed hydrogen bond formation mainly with the residues Thr436, His584, and Arg868. Notably His584 is the catalytic residue and is found to be conserved. The hydrophobic contacts are driven by Ala524, Leu490, Leu872, Ile695, and Ile888. The strong π-π interactions are also predicted between the ligand and FMN molecule or residue His584. Thus, favorable interactions between protein-ligand complexes were predicted.

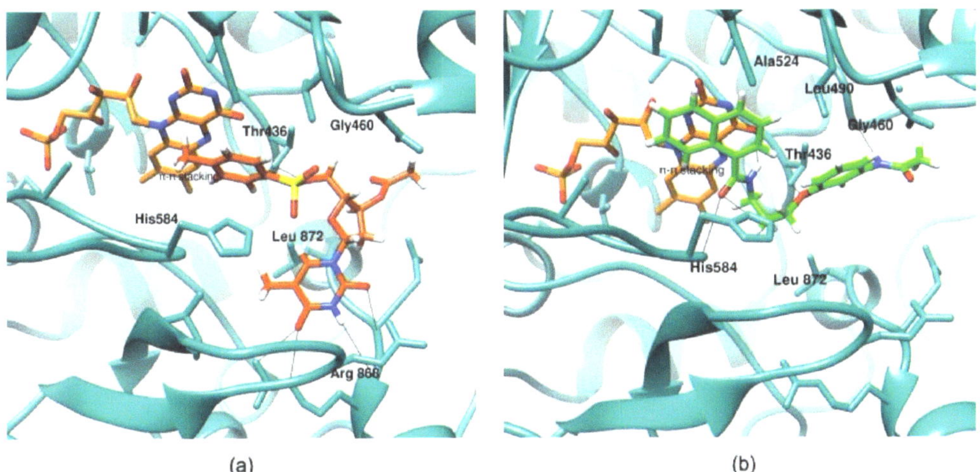

(a)

(b)

Figure 4. *Cont.*

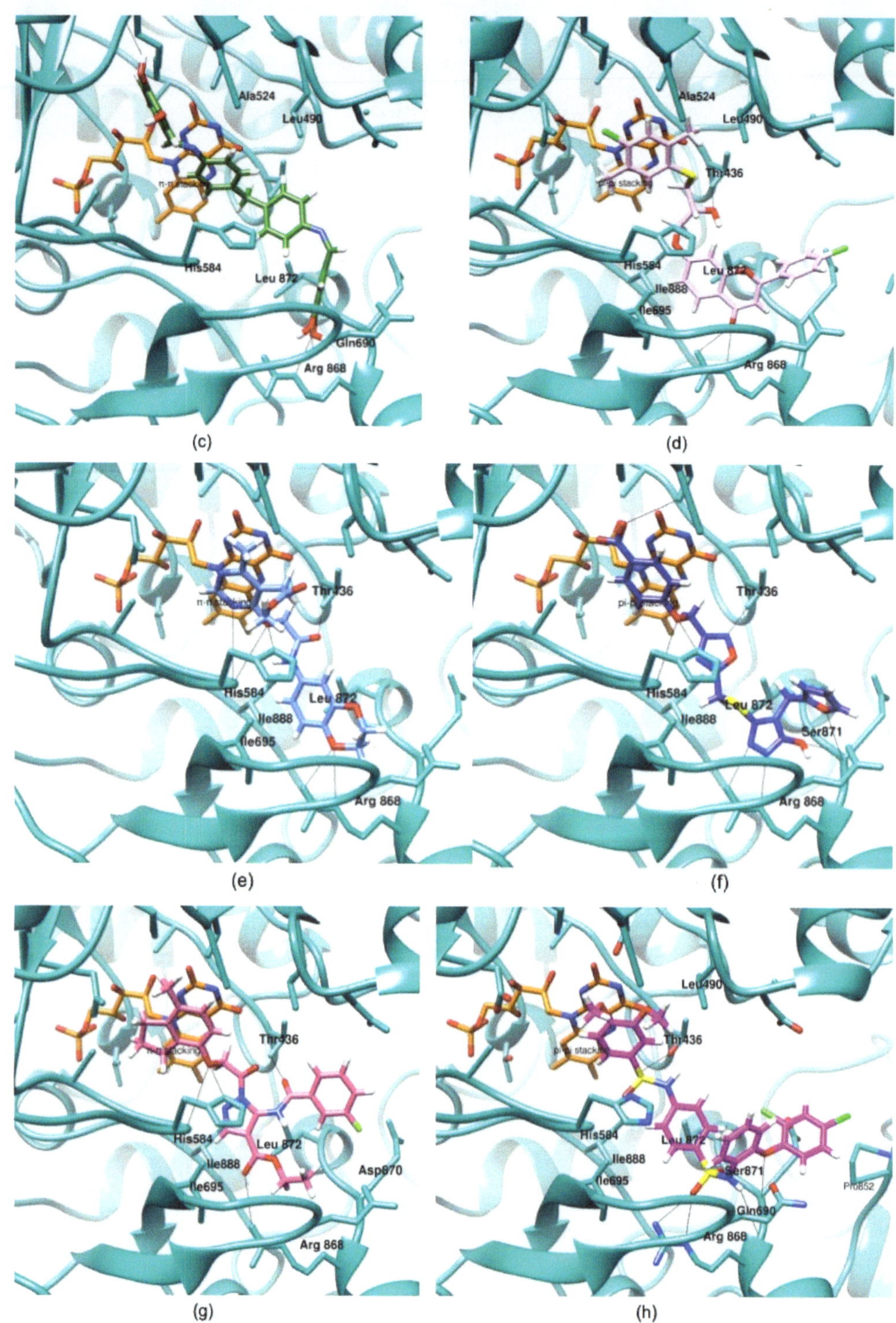

Figure 4. *Cont.*

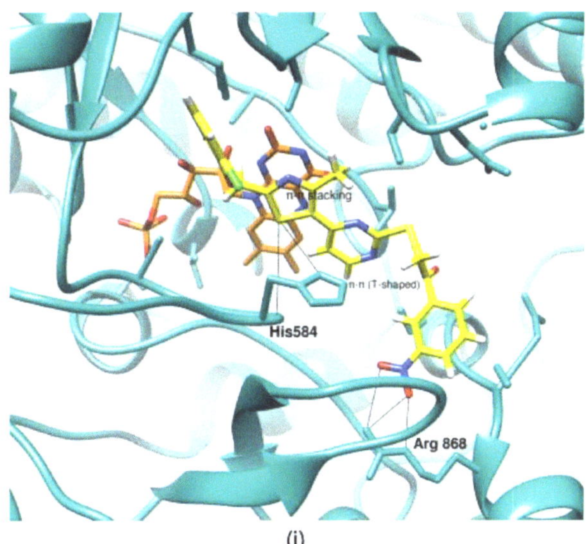

(i)

Figure 4. The 3D interaction plots for selected protein-ligand complexes (**a**) BTB14738, (**b**) BTB14516, (**c**) CD01000, (**d**) SEW02765, (**e**) HTS 07760, (**f**) HTS09453, (**g**) RH00608, (**h**) RJF01717, (**i**) SPB02705.

3.4. Molecular Dynamics Simulation

In order to obtain a better insight as well as to assess the stability of the binding mode of molecules, we conducted 10 ns long molecular dynamic simulations for all the nine proteins–ligand complexes selected after the visual inspection step. The poses selected from the Geom X docking experiment were used as the starting poses for simulation studies. Keeping in consideration the large size of FAS-I, simulation was done for the enoyl reductase domain (amino acid residue 394–1107). The NADPH binding site of the ER domain is far away from the interaction sites between the ER domain and other neighbouring domains; it is thus unlikely that the neighbouring domains contribute to the interaction between the ER domain and the proposed inhibitors. The structure of the ER domain is close to the one in the FAS-I complex, as indicated by the small RMSD in all simulations. In the next step, the resultant trajectory was analyzed for assessing the stability of the predicted protein-ligand complexes. The r.m.s.d of the ligand was calculated using g_rms command in GROMACS. The r.m.s.d for ligands was plotted for all complexes from the end of equilibration phase and was found to be low, indicating the stability of ligand poses. The average r.m.s.d of ligand atoms was calculated to 0.24, 0.18, 0.12, 0.15, 0.21, 0.11, 0.16, 0.20, and 0.21 nm for BTB14516, BTB14738, CD01000, HTS07760, HTS09453, RH00608, RJF01717, SEW02765, and SPB02705, respectively (Figure 5). The FMN molecule was found to be very stable with average r.m.s.d calculated in the range of 0.05–0.11 nm (Figure S2). The average r.m.s.d for backbone atoms of complexes was calculated in the range of 0.19–0.26 nm (Figure S2). Overall, the protein-ligand complexes were found to be significantly stable. Subsequently, the last five ns MD trajectory was used for performing MM/PBSA-based binding free energy calculations and is discussed in the next section.

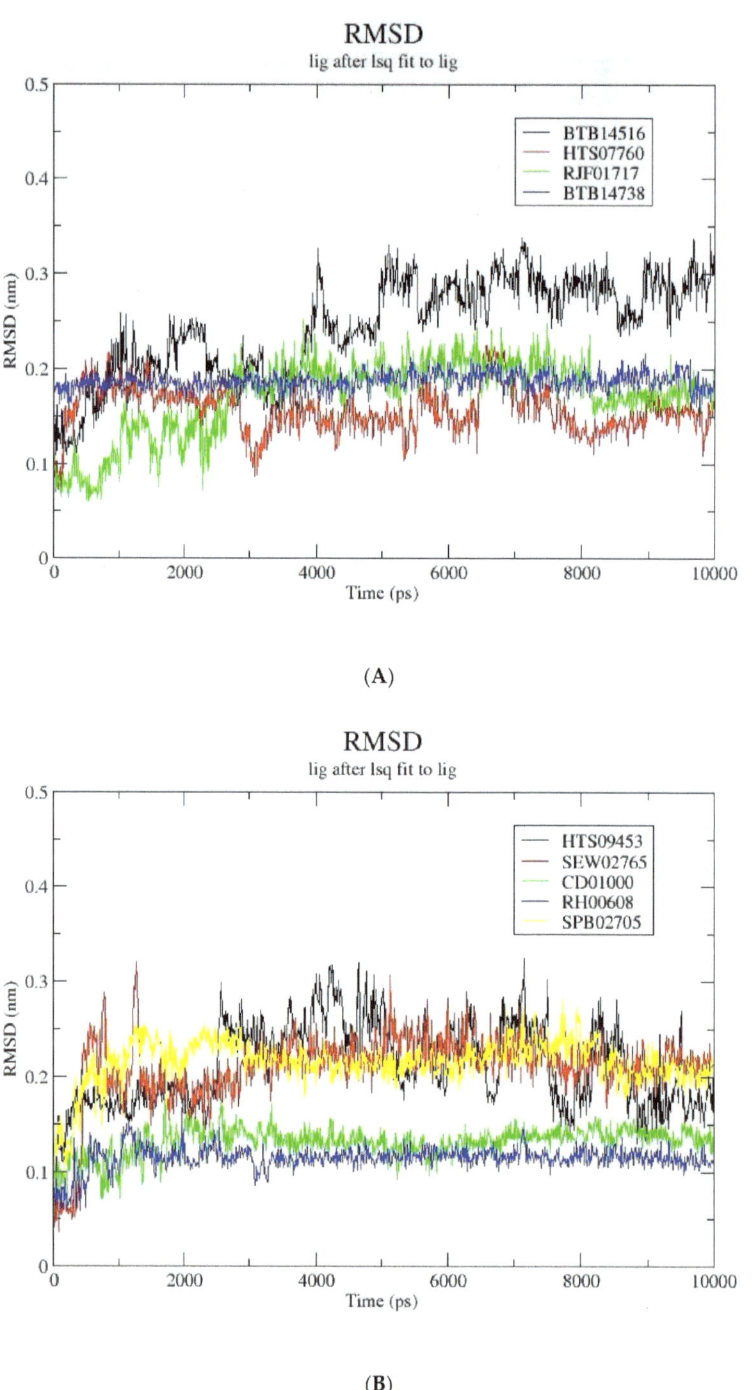

Figure 5. The r.m.s.d plots for ligand are given for (**A**) BTB14516, HTS07760, RJF01717, BTB14738 (**B**) CD01000, HTS09453, RH00608, SPB02705, and SEW02765.

3.5. MM/PBSA-Based Interaction Energy Calculation

The computational efficiency and accuracy of simulation-based end point methods such as molecular mechanics with Poisson–Boltzmann (MM/PBSA) and molecular mechanics with generalized Born and surface area (MM/GBSA) provides a reliable choice for binding free energy calculations [42]. These methods are based upon sampling of the final state of the system and the solvent is treated implicitly, thus reducing the computational time. However, these methods possess limitations regarding the estimation of conformational and solvation entropies [48]. In order to prioritize and re-rank the compounds, MM/PBSA-based binding free energy was calculated for all the nine complexes. For each system, 100 snapshots were fetched at the interval of 50 ps along the last five ns trajectory. All the compounds showed favorable binding free energy values. The compound BTB14738 was predicted to show the highest binding free energy value of -72.27 ± 12.63 KJ/mol. The other compounds RH00608, SPB02705, CD01000, and HTS07760 are also predicted to show high binding free energy values of -68.06 ± 11.80, -63.57 ± 12.22, -51.28 ± 13.74, and -53.17 ± 12.68 KJ/mol. However, the compound SEW02765 is predicted to show the lowest binding free energy value. The binding free energies of protein-ligand complexes is summarized in Table 2. The compounds were ranked upon the basis of predicted binding free energy. The selected compounds showed high predicted binding free energy, indicating better binding and potentially better inhibitory efficiency. The MM/PBSA approach has been reported to be applied reliably for re-scoring the protein-ligand complexes predicted by molecular docking. Thus, it helps in boosting the virtual screening hit rates [31,49]. The accuracy of the method is limited by the lack of conformational entropy, missing effect of water molecules in the binding site and details in the method, such as dielectric constant, continuum-solvation method, and charges [50]. Therefore, this method has been applied to systems with a varying degree of success. Nevertheless, the method has been useful to corroborate the docking results and understand the observed affinities [51,52].

Table 2. The predicted binding free energy and molecular docking score for selected protein-ligand complexes is listed.

Serial Number	Compound ID	Predicted Binding Free Energy (KJ/mol)	Molecular Docking Score
1.	BTB14738	-72.27 ± 12.63	9.55
2.	RH00608	-68.06 ± 11.80	10.80
3.	SPB02705	-63.57 ± 12.22	9.30
4.	HTS07760	-53.17 ± 12.68	9.28
5.	CD01000	-51.28 ± 13.74	10.41
6.	BTB14516	-48.91 ± 11.37	10.01
7.	RJF01717	-44.82 ± 15.77	10.18
8.	HTS09453	-43.30 ± 14.27	10.46
9.	SEW02765	-16.44 ± 13.22	9.75

Taking into consideration the predicted binding free energy values, stability of the bound ligand along with molecular docking score, four compounds (BTB14738, SPB02705, RH00608, and CD01000) were selected as potential hits (Figure 3). For compound BTB14738, the highest binding free energy of -72.27 ± 12.63 was estimated. The ligand r.m.s.d was calculated to be 0.11 nm, indicating high stability. Similarly, the compounds RH00608, SPB02705, and CD01000 also showed high favorable binding free energy values and stable ligand binding. Thus, molecular docking-based virtual screening together with molecular dynamics and the MM/PBSA method has resulted in the identification of pioneer putative hits against Mtb FAS-I.

3.6. Discussion

Firstly, the MM/PBSA result for the selected compounds BTB14738, RH00608, SPB02705, and CD01000 is given in detail in Table 3. It has been shown that Van der Waals interaction contributed more towards the favorable binding of compounds while the contribution of electrostatic interaction was estimated to be lower. Hence, Van der Waals forces are an important form of interaction between the ligand and the protein.

Table 3. The MM/PBSA results for the selected putative hits. All energies are in unit of KJ/mol.

Compound Code	Van der Waals Energy	Electrostatic Energy	Polar Solvation Energy	SASA Energy	Binding Free Energy
BTB14738	-165.92 ± 11.998	-33.77 ± 6.45	146.82 ± 13.80	-19.39 ± 0.97	-72.27 ± 12.63
RH00608	-219.005 ± 10.53	-14.61 ± 6.551	190.10 ± 10.54	-24.55 ± 1.22	-68.06 ± 11.80
SPB02705	-183.88 ± 16.99	-2.59 ± 10.16	145.40 ± 21.44	-22.50 ± 1.59	-63.57 ± 12.22
CD01000	-201.98 ± 15.30	-35.31 ± 9.28	209.04 ± 20.93	-23.04 ± 1.60	-51.28 ± 13.74

It is noteworthy that enoyl-acyl carrier protein reductases have been reported as attractive targets for the development of novel antibiotics. In one such study, the crystal structure of enoyl–ACP reductase (FabK) from Streptococcus pneumoniae in complex with phenylimidazole derivative inhibitor has been reported [53] (Figure 3). This implies that the enoyl reductase domain of FAS-I can be a promising target because it is crucial in the regulation of the pathway. The inhibitor has been reported to bind competitively with respect to NADH. The thiazole ring and a part of ureido moeity is involved in the π-π stacking interaction with the isoalloxazine ring of the FMN molecule. Similarly, the active site of Mtb FAS-I consists of the tightly bound FMN molecule and the catalytic residue His584; hence, ligands containing aromatic rings possess potential for strong π-π interactions with the FMN molecule. The presence of residues Thr436, His584, and Arg868 facilitates the formation of a hydrogen bond. Furthermore, the active site is large and open in nature, offering wide scope for optimization of compounds through additional groups. The proposed hit compounds can be pioneer inhibitors of Mtb FAS-I.

4. Conclusions

The key role of FAS-I in the survival and growth of mycobacterium makes it an attractive drug target. Only the pyrazinamide analogs have been reported as competitive inhibitors of FAS-I for NADPH binding. Therefore, in pursuit of finding potential inhibitors against FAS-I, we carried out structure-based virtual screening, focusing on the NADPH binding site of the enoyl reductase domain. Subsequently, molecular dynamics simulations were done to assess the stability of predicted binding poses and to perform MM/PBSA-based binding free energy. Based upon the predicted binding free energy values and stability of the compounds in the binding pocket, the compounds BTB14738, RH00608, SPB02705, and CD01000 have been proposed as putative hits. The calculated binding free energy indicates significant binding of the selected compounds. The proposed compounds can serve as pioneer inhibitors against Mtb FAS-I, which could pave the way for the development of a novel treatment for TB.

Supplementary Materials: The following are available online at https://www.mdpi.com/article/10.3390/app11156977/s1, Figure S1: 2D interaction plots for selected protein-ligand complexes. Figure S2: The r.m.s.d plots shown for FMN molecule and protein backbone atoms.

Author Contributions: Conceptualization, W.L.; formal analysis, N.S.; investigation, N.S.; resources, W.L.; data curation, N.S.; writing—original draft preparation, N.S.; writing—review and editing, S.-Q.M. and W.L.; visualization, S.-Q.M.; supervision, W.L.; funding acquisition, W.L. All authors have read and agreed to the published version of the manuscript.

Funding: This research was funded by Natural Science Foundation of Guangdong Province, China (Grant No. 2020A1515010984) and the Start-up Grant for Young Scientists (860-000002110384), Shenzhen University. The APC was funded by the Start-up Grant for Young Scientists (860-000002110384).

Institutional Review Board Statement: Not applicable.

Informed Consent Statement: Not applicable.

Data Availability Statement: The data presented in this study are available in the article and supplementary material.

Conflicts of Interest: The authors declare no conflict of interest.

References

1. WHO. Factsheet. 2019. Available online: https://www.who.int/tb/publications/factsheet_global.pdf?ua=1 (accessed on 22 October 2020).
2. WHO. News. 2019. Available online: https://www.who.int/news-room/fact-sheets/detail/tuberculosis (accessed on 20 October 2020).
3. Glickman, M.S.; Cox, J.S.; Jacobs, W.R. A Novel Mycolic Acid Cyclopropane Synthetase Is Required for Cording, Persistence, and Virulence of Mycobacterium tuberculosis. *Mol. Cell* **2000**, *5*, 717–727. [CrossRef]
4. Barkan, D.; Liu, Z.; Sacchettini, J.C.; Glickman, M.S. Mycolic Acid Cyclopropanation is Essential for Viability, Drug Resistance, and Cell Wall Integrity of Mycobacterium tuberculosis. *Chem. Biol.* **2009**, *16*, 499–509. [CrossRef] [PubMed]
5. Nataraj, V.; Varela, C.; Javid, A.; Singh, A.; Besra, G.S.; Bhatt, A. Mycolic acids: Deciphering and targeting the Achilles' heel of the tubercle bacillus. *Mol. Microbiol.* **2015**, *98*, 7–16. [CrossRef] [PubMed]
6. Takayama, K.; Wang, C.; Besra, G.S. Pathway to synthesis and processing of mycolic acids in Mycobacterium tuberculo-sis. *Clin. Microbiol. Rev.* **2005**, *18*, 81–101. [CrossRef] [PubMed]
7. Bhatt, A.; Molle, V.; Besra, G.S.; Jacobs, W.R., Jr.; Kremer, L. The Mycobacterium tuberculosis FAS-II condensing enzymes: Their role in mycolic acid biosynthesis, ac-id-fastness, pathogenesis and in future drug development. *Mol. Microbiol.* **2007**, *64*, 1442–1454. [CrossRef] [PubMed]
8. Brennan, J.P.; Nikaido, H. The envelope of mycobacteria. *Annu. Rev. Biochem.* **1995**, *64*, 29–63. [CrossRef] [PubMed]
9. Schweizer, E.; Hofmann, J. Microbial type I fatty acid synthases (FAS): Major players in a network of cellular FAS sys-tems. *Microbiol. Mol. Biol. Rev.* **2004**, *68*, 501–517. [CrossRef] [PubMed]
10. Lamichhane, G.; Zignol, M.; Blades, N.J.; Geiman, D.E.; Dougherty, A.; Grosset, J.; Broman, K.W.; Bishai, W.R. A postgenomic method for predicting essential genes at subsaturation levels of mutagenesis: Applica-tion to Mycobacterium tuberculosis. *Proc. Natl. Acad. Sci. USA* **2003**, *100*, 7213–7218. [CrossRef]
11. Sassetti, C.M.; Boyd, D.H.; Rubin, E.J. Comprehensive identification of conditionally essential genes in mycobacteria. *Proc. Natl. Acad. Sci. USA* **2001**, *98*, 12712–12717. [CrossRef]
12. Ma, Z.; Lienhardt, C.; McIlleron, H.; Nunn, A.; Wang, X. Global tuberculosis drug development pipeline: The need and the reality. *Lancet* **2010**, *375*, 2100–2109. [CrossRef]
13. Steele, M.A.; Prez, R.M.D. The Role of Pyrazinamide in Tuberculosis Chemotherapy. *Chest* **1988**, *94*, 845–850. [CrossRef]
14. Sayahi, H.; Pugliese, K.M.; Zimhony, O.; Jacobs, W.R.; Shekhtman, A.; Welch, J.T. Analogs of the Antituberculous Agent Pyrazinamide Are Competitive Inhibitors of NADPH Binding to M. tuberculosis Fatty Acid Synthase I. *Chem. Biodivers.* **2012**, *9*, 2582–2596. [CrossRef]
15. Zimhony, O.; Cox, J.S.; Welch, J.T.; Vilchèze, C.; Jacobs, W.R., Jr. Pyrazinamide inhibits the eukaryotic-like fatty acid synthetase I (FASI) of Mycobacterium tuberculosis. *Nat. Med.* **2000**, *6*, 1043–1047. [CrossRef]
16. Elad, N.; Baron, S.; Peleg, Y.; Albeck, S.; Grunwald, J.; Raviv, G.; Shakked, Z.; Zimhony, O.; Diskin, R. Structure of Type-I Mycobacterium tuberculosis fatty acid synthase at 3.3 Å resolution. *Nat. Commun.* **2018**, *9*, 3886. [CrossRef]
17. Maia, E.H.B.; Assis, L.C.; De Oliveira, T.A.; Da Silva, A.M.; Taranto, A.G. Structure-Based Virtual Screening: From Classical to Artificial Intelligence. *Front. Chem.* **2020**, *8*, 343. [CrossRef] [PubMed]
18. Singh, N.; Tiwari, S.; Srivastava, K.K.; Siddiqi, M.I. Identification of Novel Inhibitors of Mycobacterium tuberculosis PknG Using Pharmacophore Based Virtual Screening, Docking, Molecular Dynamics Simulation, and Their Biological Evaluation. *J. Chem. Inf. Model.* **2015**, *55*, 1120–1129. [CrossRef] [PubMed]
19. Kumar, A.; Siddiqi, M.I.; Miertus, S. New molecular scaffolds for the design of Mycobacterium tuberculosis type II dehy-droquinase inhibitors identified using ligand and receptor based virtual screening. *J. Mol. Model.* **2010**, *16*, 693–712. [CrossRef] [PubMed]
20. Vilar, S.; Sobarzo-Sanchez, E.; Santana, L.; Uriarte, E. Molecular Docking and Drug Discovery in β-Adrenergic Receptors. *Curr. Med. Chem.* **2017**, *24*, 4340–4359. [CrossRef] [PubMed]
21. Lionta, E.; Spyrou, G.; Vassilatis, D.K.; Cournia, Z. Structure-Based Virtual Screening for Drug Discovery: Principles, Applications and Recent Advances. *Curr. Top. Med. Chem.* **2014**, *14*, 1923–1938. [CrossRef] [PubMed]
22. Wang, T.; Wu, M.-B.; Chen, Z.-J.; Chen, H.; Lin, J.-P.; Yang, L.-R. Fragment-based drug discovery and molecular docking in drug design. *Curr. Pharm. Biotechnol.* **2015**, *16*, 11–25. [CrossRef]

23. Moro, W.B.; Yang, Z.; Kane, T.A.; Brouillette, C.G.; Brouillette, W.J. Virtual screening to identify lead inhibitors for bacterial NAD synthetase (NADs). *Bioorg. Med. Chem. Lett.* **2009**, *19*, 2001–2005. [CrossRef]

24. Mishra, A.K.; Singh, N.; Agnihotri, P.; Mishra, S.; Singh, S.P.; Kolli, B.K.; Chang, K.P.; Sahasrabuddhe, A.A.; Siddiqi, M.I.; Pratap, J.V. Discovery of novel inhibitors for Leishmania nucleoside diphosphatase kinase (NDK) based on its struc-tural and functional characterization. *J. Comput. Aided. Mol. Des.* **2017**, *31*, 547–562. [CrossRef]

25. Lee, Y.-V.; Choi, S.B.; Wahab, H.A.; Lim, T.S.; Choong, Y.S. Applications of Ensemble Docking in Potential Inhibitor Screening forMycobacterium tuberculosisIsocitrate Lyase Using a Local Plant Database. *J. Chem. Inf. Model.* **2019**, *59*, 2487–2495. [CrossRef]

26. Kwofie, S.K.; Adobor, C.; Quansah, E.; Bentil, J.; Ampadu, M.; Miller, W.A.; Wilson, M.D. Molecular docking and dynamics simu-lations studies of OmpATb identifies four potential novel natural product-derived anti-Mycobacterium tuberculosis compounds. *Comput. Biol. Med.* **2020**, *122*, 103811. [CrossRef]

27. Zhao, W.; Xiong, M.; Yuan, X.; Li, M.; Sun, H.; Xu, Y. In Silico Screening-Based Discovery of Novel Inhibitors of Human Cyclic GMP–AMP Synthase: A Cross-Validation Study of Molecular Docking and Experimental Testing. *J. Chem. Inf. Model.* **2020**, *60*, 3265–3276. [CrossRef]

28. Newton, A.S.; Faver, J.C.; Micevic, G.; Muthusamy, V.; Kudalkar, S.N.; Bertoletti, N.; Anderson, K.S.; Bosenberg, M.W.; Jorgensen, W.L. Structure-Guided Identification of DNMT3B Inhibitors. *ACS Med. Chem. Lett.* **2020**, *11*, 971–976. [CrossRef] [PubMed]

29. Vázquez-Jiménez, L.K.; Paz-González, A.D.; Juárez-Saldivar, A.; Uhrig, M.L.; Agusti, R.; Reyes-Arellano, A.; Nogueda-Torres, B.; Rivera, G. Structure-Based Virtual Screening of New Benzoic Acid Derivatives as Trypanosoma cruzi Trans-sialidase Inhibitors. *Med. Chem.* **2020**, *16*, 1–9. [CrossRef]

30. Gupta, D.; Singh, A.; Somvanshi, P.; Singh, A.; Khan, A.U. Structure-Based Screening of Non-β-Lactam Inhibitors against Class D β-Lactamases: An Approach of Docking and Molecular Dynamics. *ACS Omega* **2020**, *5*, 9356–9365. [CrossRef] [PubMed]

31. Poli, G.; Granchi, C.; Rizzolio, F.; Tuccinardi, T. Application of MM-PBSA Methods in Virtual Screening. *Molecules* **2020**, *25*, 1971. [CrossRef] [PubMed]

32. Maybridge Library. Available online: http://www.maybridge.com/ (accessed on 25 December 2020).

33. Jain, A.N. Surflex: Fully Automatic Flexible Molecular Docking Using a Molecular Similarity-Based Search Engine. *J. Med. Chem.* **2003**, *46*, 499–511. [CrossRef] [PubMed]

34. Berendsen, H.J.; van der Spoel, D.; van Drunen, R. GROMACS—A message-passing parallel molecu-lar-dynamics implementation. *Comput. Phys. Commun.* **1995**, *91*, 43–56. [CrossRef]

35. Vanommeslaeghe, K.; Hatcher, E.; Acharya, C.; Kundu, S.; Zhong, S.; Shim, J.; Darian, E.; Guvench, O.; Lopes, P.E.M.; Vorobyov, I.; et al. CHARMM general force field: A force field for drug-like molecules compatible with the CHARMM all-atom additive biological force fields. *J. Comput. Chem.* **2009**, *31*, 671–690. [CrossRef] [PubMed]

36. Vanommeslaeghe, K.; Raman, E.P.; MacKerell, A.D. Automation of the CHARMM General Force Field (CGenFF) II: As-signment of bonded parameters and partial atomic charges. *J. Chem. Inf. Model.* **2012**, *52*, 3155–3168. [CrossRef]

37. Darden, T.; York, D.; Pedersen, L. Particle mesh Ewald: An N·log(N) method for Ewald sums in large systems. *J. Chem. Phys.* **1993**, *98*, 10089–10092. [CrossRef]

38. Available online: http://plasma-gate.weizmann.ac.il/Grace/ (accessed on 12 April 2020).

39. Hou, T.; Wang, J.; Li, Y.; Wang, W. Assessing the performance of the molecular mechanics/Poisson Boltzmann surface area and molecular me-chanics/generalized Born surface area methods. II. The accuracy of ranking poses generated from docking. *J. Comput. Chem.* **2011**, *32*, 866–877. [CrossRef]

40. Kumari, R.; Kumar, R.; Open Source Drug Discovery Consortium; Lynn, A. g_mmpbsa—A GROMACS tool for high-throughput MM/PBSA calculations. *J. Chem. Inf. Model.* **2014**, *54*, 1951–1962. [CrossRef] [PubMed]

41. Wright, D.; Hall, B.A.; Kenway, O.A.; Jha, S.; Coveney, P.V. Computing Clinically Relevant Binding Free Energies of HIV-1 Protease Inhibitors. *J. Chem. Theory Comput.* **2014**, *10*, 1228–1241. [CrossRef]

42. Xu, L.; Sun, H.; Li, Y.; Wang, J.; Hou, T. Assessing the Performance of MM/PBSA and MM/GBSA Methods. The Impact of Force Fields and Ligand Charge Models. *J. Phys. Chem. B* **2013**, *117*, 8408–8421. [CrossRef]

43. Homeyer, N.; Gohlke, H. Free Energy Calculations by the Molecular Mechanics Poisson–Boltzmann Surface Area Method. *Mol. Inform.* **2012**, *31*, 114–122. [CrossRef]

44. Baker, N.A.; Sept, D.; Joseph, S.; Holst, M.J.; McCammon, J.A. Electrostatics of nanosystems: Application to microtubules and the ribosome. *Proc. Natl. Acad. Sci. USA* **2001**, *98*, 10037–10041. [CrossRef]

45. Levy, R.M.; Zhang, L.Y.; Gallicchio, E.; Felts, A.K. On the nonpolar hydration free energy of proteins: Surface area and continuum solvent models for the so-lute-solvent interaction energy. *J. Am. Chem. Soc.* **2003**, *125*, 9523–9530. [CrossRef]

46. Tan, C.; Tan, Y.-H.; Luo, R. Implicit Nonpolar Solvent Models. *J. Phys. Chem. B* **2007**, *111*, 12263–12274. [CrossRef]

47. Jenni, S.; Leibundgut, M.; Boehringer, D.; Frick, C.; Mikolásek, B.; Ban, N. Structure of Fungal Fatty Acid Synthase and Implications for Iterative Substrate Shuttling. *Science* **2007**, *316*, 254–261. [CrossRef] [PubMed]

48. Wang, E.; Sun, H.; Wang, J.; Wang, Z.; Liu, H.; Zhang, J.Z.; Hou, T. End-Point Binding Free Energy Calculation with MM/PBSA and MM/GBSA: Strategies and Applications in Drug Design. *Chem. Rev.* **2019**, *119*, 9478–9508. [CrossRef]

49. Botelho, F.D.; Gonçalves, A.S.; França, T.C.; LaPlante, S.R.; de Almeida, J.S. Identification of novel potential ricin inhibitors by virtual screening, molecular docking, molecular dy-namics and MM/PBSA calculations: A drug repurposing approach. *J. Biomol. Struct. Dyn.* **2021**. [CrossRef]

50. Genheden, S.; Ryde, U. The MM/PBSA and MM/GBSA methods to estimate ligand-binding affinities. *Expert Opin. Drug Discov.* **2015**, *10*, 449–461. [CrossRef] [PubMed]
51. Laurini, E.; Col, V.D.; Mamolo, M.G.; Zampieri, D.; Posocco, P.; Fermeglia, M.; Vio, L.; Pricl, S. Homology Model and Docking-Based Virtual Screening for Ligands of the σ1 Receptor. *ACS Med. Chem. Lett.* **2011**, *2*, 834–839. [CrossRef] [PubMed]
52. Venken, T.; Krnavek, D.; Münch, J.; Kirchhoff, F.; Henklein, P.; De Maeyer, M.; Voet, A. An optimized MM/PBSA virtual screening approach applied to an HIV-1 gp41 fusion peptide inhibitor. *Proteins Struct. Funct. Bioinform.* **2011**, *79*, 3221–3235. [CrossRef]
53. Saito, J.; Yamada, M.; Watanabe, T.; Iida, M.; Kitagawa, H.; Takahata, S.; Ozawa, T.; Takeuchi, Y.; Ohsawa, F. Crystal structure of enoyl-acyl carrier protein reductase (FabK) from *Streptococcus pneumoniae* reveals the binding mode of an inhibitor. *Protein. Sci.* **2008**, *17*, 691–699. [CrossRef]

 *applied sciences*

Article

Protein Fluctuations in Response to Random External Forces

Domenico Scaramozzino [1],*, Pranav M. Khade [2],* and Robert L. Jernigan [2],*

[1] Department of Structural, Geotechnical and Building Engineering, Politecnico di Torino, Corso Duca degli Abruzzi 24, 10129 Torino, Italy

[2] Roy J. Carver Department of Biochemistry, Biophysics and Molecular Biology, Iowa State University, Ames, IA 50011, USA

* Correspondence: domenico.scaramozzino@polito.it (D.S.); pranavk@iastate.edu (P.M.K.); jernigan@iastate.edu (R.L.J.)

Abstract: Elastic network models (ENMs) have been widely used in the last decades to investigate protein motions and dynamics. There the intrinsic fluctuations based on the isolated structures are obtained from the normal modes of these elastic networks, and they generally show good agreement with the B-factors extracted from X-ray crystallographic experiments, which are commonly considered to be indicators of protein flexibility. In this paper, we propose a new approach to analyze protein fluctuations and flexibility, which has a more appropriate physical basis. It is based on the application of random forces to the protein ENM to simulate the effects of collisions of solvent on a protein structure. For this purpose, we consider both the C^α-atom coarse-grained anisotropic network model (ANM) and an elastic network augmented with points included for the crystallized waters. We apply random forces to these protein networks everywhere, as well as only on the protein surface alone. Despite the randomness of the directions of the applied perturbations, the computed average displacements of the protein network show a remarkably good agreement with the experimental B-factors. In particular, for our set of 919 protein structures, we find that the highest correlation with the B-factors is obtained when applying forces to the external surface of the water-augmented ANM (an overall gain of 3% in the Pearson's coefficient for the entire dataset, with improvements up to 30% for individual proteins), rather than when evaluating the fluctuations obtained from the normal modes of a standard C^α-atom coarse-grained ANM. It follows that protein fluctuations should be considered not just as the intrinsic fluctuations of the internal dynamics, but also equally well as responses to external solvent forces, or as a combination of both.

Keywords: elastic network model; protein flexibility; B-factors; protein fluctuations; random force application; protein surface

Citation: Scaramozzino, D.; Khade, P.M.; Jernigan, R.L. Protein Fluctuations in Response to Random External Forces. *Appl. Sci.* **2022**, *12*, 2344. https://doi.org/10.3390/app12052344

Academic Editor: Hervé Quiquampoix

Received: 1 December 2021
Accepted: 19 February 2022
Published: 23 February 2022

Publisher's Note: MDPI stays neutral with regard to jurisdictional claims in published maps and institutional affiliations.

1. Introduction

The B-factors of a protein, the Debye-Waller factors or temperature factors, are measures of the atomic displacements about their equilibrium position, i.e., atomic fluctuations [1–3], but also the effects of multiple conformations as well as errors in the structures. They are generally accepted to be mostly the result of internal protein dynamics and any static disorder [4]. They have also been shown to be associated with protein flexibility and to correspond closely to protein mechanisms [5–9]. B-factors have been associated with protein flexibility, which is strictly related to protein action and function [10–12]. The experimental B-factors obtained from X-ray crystallography have been reproduced fairly accurately by various computational models.

One of the most widely used computational methods for investigating protein dynamics and fluctuations has been molecular dynamics (MD). MD simulations have proven their usefulness for investigations of protein folding, enzyme catalysis, and protein mechanisms in general [13–15]. Also, it has been shown that the MD-derived atomic fluctuations due to the internal protein motions show some degree of agreement with the experimental

B-factors [16,17]. However, due to the high computational burden of MD simulations, these can sometimes be expensive for investigating the large molecular complexes, especially regarding the slowest protein motions, accessible only at long simulation times. These slow motions are in fact usually the ones most closely related to the functional mechanisms of the protein and can take place on longer time scales than may be accessible in standard MD simulations. The harmonicity assumption has been exploited for the extraction of the low-frequency protein dynamics [18–21]. Normal mode analysis (NMA) came into play as a simplified yet powerful tool to investigate the slower protein motions and for evaluating protein fluctuations and mechanisms [22,23], even in torsional space [24].

The seminal work of Tirion [25] showed that even a single-parameter harmonic potential, only based on the elastic properties of a network of Hookean springs connecting the protein atoms, was sufficient to reproduce the slow dynamics in good detail. All of the elastic network models are essentially entropic models since there is not usually any distinction of atom or amino acid types, i.e., all springs are taken to be similar in character. A further step towards simplification came with the coarse-graining development for these elastic network models (ENMs). Among the ENMs, the gaussian network model (GNM) was developed to obtain insights into protein dynamics and fluctuations simply by diagonalizing the Kirchhoff matrix, built by using the network contacts between close neighboring C^α atoms [26–31]. Despite the remarkable correlations obtained between the GNM-based fluctuations and the experimental B-factors, the GNM lacks the information about the directions of motions, since it assumes that the motions are fundamentally isotropic in all directions [28]. The anisotropic network model (ANM) was then developed to include the three-dimensional directionality in the calculation of protein motions [32]. The ANM was then improved by various research groups to achieve higher correlations between the computed fluctuations and the experimental B-factors [33–37]. These elastic models were subsequently used to study the conformational changes of proteins arising from sets of low-frequency modes [38–48] as well as to generate feasible pathways between two known conformations [48–54].

Structural elastic models, particularly the ANM, were applied widely for the investigation of protein dynamics, fluctuations, and mechanism. However, they are also well-suited for the analysis of the protein structural responses from the application of external perturbations. Based on the work from Ikeguchi et al. [55], who showed that protein conformational changes upon ligand-binding could be analyzed based on linear response theory, the perturbation-response scanning (PRS) method was proposed by the Atilgan group [56,57]. Randomly oriented forces were applied at selected residues, and the corresponding response of the ferric binding protein [56] and another 24 proteins [57] were found to agree fairly accurately with the experimentally detected conformational change. A similar study was conducted by Gerek and Ozkan [58] to study the allosteric network in PDZ domains. A PRS-based technique, coupled with energy-based Metropolis Monte Carlo (MMC) simulations, was carried out by Liu et al. [59] to simulate the closed-to-open conformational change of a GroEL subunit due to directional forces presumed to originate from exothermic ATP hydrolysis. Interestingly, some of the apparent conformational changes being attributed to the binding of ATP or ADP may originate from the exothermic forces generated by hydrolysis. Recently, it was also shown that the application of forces in a dynamic fashion is able to drive the conformational change with a strong directionality correlation [60]. Eyal and Bahar [61] investigated the mechanical response of protein structure to external pulling forces in order to detect the anisotropic mechanical resistance to explain the outcomes of single-molecule manipulation techniques. More recently, we made use of a similar pairwise force application methodology in order to measure the overall protein flexibility by using the engineering concepts of structural compliance and stiffness [62].

Most of the works based on the coarse-grained ENMs include only one or a few representative atoms of the amino acids in the protein network, e.g., the C^α atoms. Remarkably it has been seen that this geometric coarse-graining at the level of one point per amino acid yields almost exactly the same motions as from a full atomic elastic model. This result is believed to be the result of the dense packing leading to the strong stability of protein domains [63]. However, most of these models do not explicitly account for the protein surface, which is the part most exposed to the surrounding environment. Water and small molecules can often be tightly bound to the protein surface, thereby affecting what is actually considered to be the surface of a protein structure. The role of such tightly bound crystalized waters in protein dynamics has been studied in the last few decades [64–67]. There have also been investigations of the solvent network surrounding the protein and its effect on the dynamics [68–70]. The inclusion of water molecules in the structure yields some increases in the quality of calculated enthalpies and of the residue interaction network [71,72]. This is one of the important reasons why all-atom MD simulations usually include these explicit waters.

This paper presents a novel method based on random perturbations applied to protein ENMs to assess protein fluctuations and flexibilities. Random forces are applied both throughout the complete protein elastic network and also separately to only the protein surface, which is exposed to the surrounding environment. In addition, a water-enriched ENM is considered, where the water molecules whose coordinates are given in the Protein Data Bank (PDB) files [73] are used as additional nodes in the elastic network. These latter force application simulations aim to mimic the random collisions occurring on the protein structure due to the interaction with the solvent and other solutes. From the calculation of the displacements within the protein network, i.e., the protein responses, we show that a good correlation is found with the experimental B-factors, thus leading to a good prediction of the protein flexibilities. The correlations with the usual mode-based fluctuations are also reported for comparison. It is also found that, in most cases, applying random forces on the surface of the water-enriched protein ENM leads to the highest correlation between the resulting displacements and the experimental B-factors. This demonstrates that the protein fluctuations may reflect more than the internal dynamics alone, and also include some effects from the continuous random bombardments or restraints by the surrounding solvent on the protein structure.

2. Methods

In this section, we briefly recount the fundamentals of the Anisotropic Network Model (ANM) [32], that is commonly used for generating the fluctuations in terms of the normal modes, and then we describe the computational framework related to the presently adopted force applications on the elastic networks.

2.1. Anisotropic Network Model (ANM) and the Calculation of Normal Mode-Based Fluctuations

ANM relies on the assumption that proteins can be modeled as simple elastic networks, made up of point nodes connected by linear elastic springs, allowing insights regarding fluctuations and global mechanisms [32,34,74]. For a system of N points, e.g., N residues in the one-bead-per-residue coarse-grained representation, the $3N \times 3N$ Hessian matrix of the system takes the following form:

$$
H = \begin{bmatrix}
H_{1,1} & \cdots & H_{1,i} & \cdots & H_{1,j} & \cdots & H_{1,N} \\
\cdots & \cdots & \cdots & \cdots & \cdots & \cdots & \cdots \\
H_{i,1} & \cdots & H_{i,i} & \cdots & H_{i,j} & \cdots & H_{i,N} \\
\cdots & \cdots & \cdots & \cdots & \cdots & \cdots & \cdots \\
H_{j,1} & \cdots & H_{j,i} & \cdots & H_{j,j} & \cdots & H_{j,N} \\
\cdots & \cdots & \cdots & \cdots & \cdots & \cdots & \cdots \\
H_{N,1} & \cdots & H_{N,i} & \cdots & H_{N,j} & \cdots & H_{N,N}
\end{bmatrix},
\tag{1}
$$

where each 3×3 submatrix $H_{i,j}$ contains the stiffness coefficients of the springs connecting nodes i and j. The off-diagonal submatrix $H_{i,j}$ is computed based on the harmonic potential of the elastic spring with force constant $\gamma_{i,j}$. The diagonal submatrices $H_{i,i}$ are calculated as the summation involving all the nodes linked to ith node with a negative sign [32]. The model, and consequently the Hessian matrix, depends on some numerical parameters: the usual model uses a cut-off limit in distance to define the network topology. The original ANM was developed by considering equal spring constants for all connections, i.e., $\gamma_{i,j} = \gamma$, and a geometrical cut-off r_c was applied in order to consider springs placed only between close nodes. Typical values of r_c in the ANM are around 15 Å. Later on, distance-dependent force constants were introduced [33,34], as:

$$\gamma_{i,j} \propto \frac{1}{(r_{i,j}{}^0)^p},$$ (2)

where p represents an inverse number for the decay parameter that allows connecting all points in a structure, with springs with variable strength—lower spring constants for longer inter-node distances $r_{i,j}{}^0$. This distance-dependent spring network was shown to provide an improved agreement between the results and experimental data [33,34].

Once the Hessian matrix is computed based on the protein coordinates from the PDB file [73] and the spring connectivity, the $3N$ eigenvalues λ_n and $3N$ eigenvectors U_n are obtained by solving the eigenvalue-eigenvectors decomposition. Due to the fact that the protein structure is usually not externally constrained, the first six eigenvalues are found to be zero, with corresponding mode shapes associated with the six rigid-body motions of the whole molecule. Therefore, these six motions are factored out and singular value decomposition is used to obtain the normal modes. Hence, the fluctuations based on the normal modes can be easily calculated as [22,75,76]:

$$B_i = \frac{8}{3}\pi^2 k_B T \sum_{n=7}^{3N} \frac{U_{i,n}{}^2}{\lambda_n},$$ (3)

where B_i represents the computed B-factor for residue i, k_B is the Boltzmann constant, T is the absolute temperature, $U_{i,n}$ stands for the displacement of node i in the nth mode, and λ_n is the eigenvalue for the nth mode.

2.2. Force Application on Elastic Networks

Here, we propose a new approach for the calculation of protein flexibility and fluctuations. This approach is based on the application of random forces on protein elastic networks. These forces are intended to simulate the external perturbations that arise from the external environment, i.e., protein-solvent interactions, Brownian motions, collisions of molecules, etc. The reality, however, is that the environment is not usually known, but cryoEM has the promise of providing some information about this.

In this work, we use two different ENMs for modelling the protein structure. The first one is the parameter-free anisotropic network model (pfANM) [33], where the C^α atoms are the only nodes used to build the protein network and all the C^α-C^α connections are considered to be linked with distance-dependent springs. In the second model, the water molecules contained in the PDB file are also added to the network as additional nodes. Additional springs are correspondingly created that connect the water molecules to all the other nodes of the network. We refer to this second model as the water-pfANM (wpfANM). Both models are built by considering a decay exponent p for the spring constant equal to 3 (see Equation (2)), based on results shown to yield the best results analyzed in our previous work [62].

The response of the protein structure to external perturbations is evaluated by applying forces to the nodes of the network and consequently computing the corresponding elastic displacements. Various force application patterns are considered here. For the pfANM, the perturbations are applied both to the complete structure, i.e., on all C^α atoms, and separately only to the nodes lying on the external protein surface. For the wpfANM, three different force patterns are considered: (1) forces acting on the entire network, i.e., all the C^α atoms and water molecules, (2) only on the nodes lying on the protein-water network surface, and (3) only on the water molecules.

For each of the models considered (pfANM and wpfANM) and their force application patterns, the calculation is based on the generation of a random 3×1 force vector F_i^s for each node i to be perturbed for each simulation s. The three scalar components of this force vector, i.e., $F_{i,x}^s$, $F_{i,y}^s$ and $F_{i,z}^s$, are sampled from a uniform distribution U in the interval $[-1,1]$. The complete force vector F^s is then generated by assembling all the 3×1 nodal vectors. Note that F^s is a $3N \times 1$ vector, with N the number of C^α atoms in the pfANM case or the total number of C^α atoms plus water molecules in the wpfANM. Once the force vector F^s is defined, it is straightforward to compute the $3N \times 1$ displacement vector δ^s for each simulation s that contains the elastic displacements of the nodes, i.e., the protein response, as follows:

$$\delta^s = H^{-1} F^s, \tag{4}$$

where H^{-1} is the pseudo-inverse of the elastic network. H^{-1} can be computed from the eigenvalues and eigenvectors of the Hessian matrix as:

$$\tilde{H}^{-1} = \sum_{n=7}^{3N} \frac{U_n U_n^T}{\lambda_n}. \tag{5}$$

From the displacement vector δ^s, the total displacement of each ith node can be computed as:

$$\delta_i^s = \sqrt{\left(\delta_{i,x}^s\right)^2 + \left(\delta_{i,y}^s\right)^2 + \left(\delta_{i,z}^s\right)^2}, \tag{6}$$

With $\delta_{i,x}^s$, $\delta_{i,y}^s$ and $\delta_{i,z}^s$ being the three Cartesian components of the node displacements.

This procedure is repeated multiple times in order to generate a sample with different random force vectors F^s and evaluating the corresponding node displacements δ^s each time. The average displacement of each node i is then evaluated as the average of all the displacements δ_i^s over the total number of simulations S:

$$\delta_i = \frac{1}{S} \sum_{s=1}^{S} \delta_i^s. \tag{7}$$

In this analysis, we have generated a sample of 10,000, i.e., $S = 10,000$. Then, the average displacement δ_i of the ith residue for the sample can be compared to the experimental B-factors available in the PDB file. Pearson's correlation coefficient can finally be used to estimate the similarity between the two distributions, i.e., between the experimental B-factors and the simulated average displacements of the protein network due to the random perturbations. As a result, high Pearson's coefficients would indicate a high degree of similarity between the computed protein fluctuations and the experimental B-factors.

We mention above that different force application patterns are considered in this study. Specifically, besides applying forces to the entire protein network in the pfANM and wpfANM, and to the water molecules alone in the wpfANM, we also apply forces only to the nodes lying on the external protein surface (pfANM) and on the protein-water network surface (wpfANM). The reason for this is due to the fact that the effect of random collisions is more likely to occur on the exterior protein surface, rather than in the interior. For this purpose, the surface residues were calculated by computing the boundary geometry of the set of 3D coordinates of the network points. The external nodes were defined as those lying on the boundary surface. In this analysis, the generation of this surface was dependent on a parameter, known as the shrink factor. The shrink factor characterizes the amount of shrinkage of the boundary geometry, with values ranging from 0 to 1: zero corresponds to the convex hull, one corresponds to the maximum shrunk boundary. Note that the shrink factors used here for the generation of the external surface correspond to the normalized alpha shape, recently used by us [77] to extract hinge-domain information from protein structures. By using the approach based on the shrink factor, different external surfaces are generated by varying this numerical parameter from 0 to 1, in steps of 0.1. The surfaces obtained are finally used to select the external nodes on which the external forces will be applied.

2.3. Protein Dataset and Summary of the Models and Analyses

The analysis was performed on the same dataset used in our previous work [62] that includes 921 high-resolution X-ray single-chain protein structures from the PDB. The resolution of the selected crystal structures is below 1.3 Å, with a maximum sequence identity of 30%. Two structures were removed from the dataset, i.e., 1IXH and 2BK9, because of errors found in the PDB file regarding the waters. The size of the 919 final proteins range from 101 to 1174 residues.

As mentioned in Section 2.2, the pfANM and wpfANM were built for all protein structures by neglecting and considering water molecules, respectively. For both models, after evaluating the Hessian matrix, the mode-based fluctuations were evaluated according to Equation (3) and compared to the experimental B-factors. Then, for the pfANM, random forces were applied: (1) to all the nodes of the network and (2) only to the external nodes. Similarly, for the wpfANM, forces are considered to be acting on: (1) the entire network, (2) only on the external nodes and (3) only on the waters. From each of these simulations, the nodal displacements are computed from Equation (7) and Pearson correlation coefficients with the experimental B-factors are also calculated. In Table 1, the summary of the models and simulations and their designators for their Pearson correlation coefficients are given.

Table 1. The Designators for the Pearson Collection Coefficients for the Various Elastic Network Models.

Model	pfANM			wpfANM			
Nodes in the network	C^α atoms			C^α atoms + water molecules *			
Analysis	Mode-based fluctuations	Force application		Mode-based fluctuations	Force application		
Nodes perturbed	-	All nodes	External nodes	-	All nodes	External nodes **	Water nodes ***
Correlation coefficient	ρ_{FL}	$\rho_{FR,ALL}$	$\rho_{FR,EXT}$	$\rho_{W,FL}$	$\rho_{W,FR,ALL}$	$\rho_{W,FR,EXT}$	$\rho_{W,FR,WAT}$

* All waters in the pdb are included, without checking whether they are exterior. ** All nodes lying on the external surface (boundary geometry) of the protein-water network are perturbed by random forces; *** Only water molecules in the protein-water network are perturbed by random forces.

3. Results and Discussion

3.1. Fluctuations and Force Application on the pfANM

In this section, the flexibility of the protein structure will be investigated by using traditional pfANM mode-based fluctuations as well as from the outcome of the random force applications to the protein. Figure 1 shows the correlation coefficients obtained from a comparison of the experimental B-factors with the mode-based fluctuations as well as the average displacements due to the applied random forces. Figure 1a,b report the Pearson coefficients obtained for the 919 single-chain protein structures, with the values ordered by ascending values of ρ_{FL}. Figure 1c shows the statistical distributions of the correlation coefficients, whose median values and standard deviations are reported in the keys.

Figure 1a,b shows the distributions of correlation coefficients $\rho_{FR,ALL}$ and $\rho_{FR,EXT}$, that are due to the random force applications, and these are observed to oscillate near the population of ρ_{FL}. This means that the average displacements of the protein elastic network due to the force application are indeed well correlated with the experimental B-factors, with a similar agreement as for the traditionally used mode-based fluctuations. The same conclusions can also be drawn by looking at Figure 1c, where the three statistical distributions exhibit the same pattern and similar median values, i.e., 0.63, 0.63 and 0.64, for ρ_{FL}, $\rho_{FR,ALL}$ and $\rho_{FR,EXT}$, respectively. Therefore, it cannot be concluded that applying random forces on the protein structures always enhances the correlation with the experimental B-factors, while it can be concluded that perturbing the protein structure by applying random forces leads to a good estimate of the experimentally determined fluctuations, at least as good as those found with the normal modes.

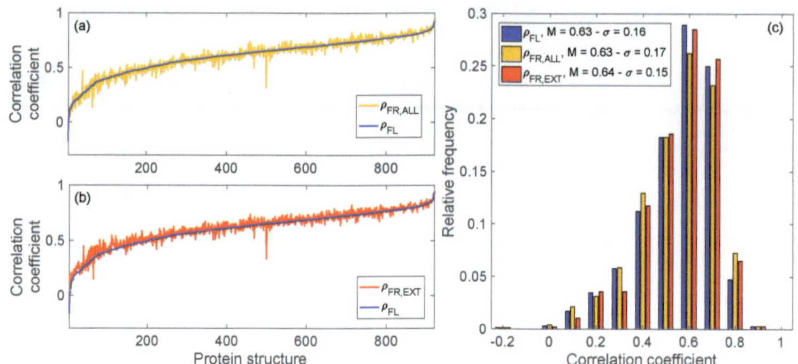

Figure 1. Correlations of experimental B-factors with ENM-based fluctuations and the average displacements due to random perturbations (pfANM): (**a,b**) correlation coefficients for the analyzed 919 protein structures (blue for ρ_{FL}, orange for $\rho_{FR,ALL}$, and red for $\rho_{FR,EXT}$); (**c**) statistical distribution of the obtained correlation coefficients. The results are all very similar, showing relatively little differences among them.

As mentioned in the previous section, the application of random forces on the external protein surface requires the selection of the nodes that lie on the exterior. For this purpose, various external boundaries were generated by changing the shrink factor of the surface, varying from 0 to 1 with steps of 0.1. Figure 2 shows an example of different external surfaces generated with shrink factors of 0, 0.2, 0.4, 0.6, 0.8 and 1 for the infrared fluorescent protein (PDB: 5AJG). As can be observed, increasing the shrink factor leads to considering a higher number of nodes lying on the surface, which in turn has a more detailed structure. Since the primary determinant of a structure's dynamics is its shape, clearly the most detailed structure would be expected to be the best [78].

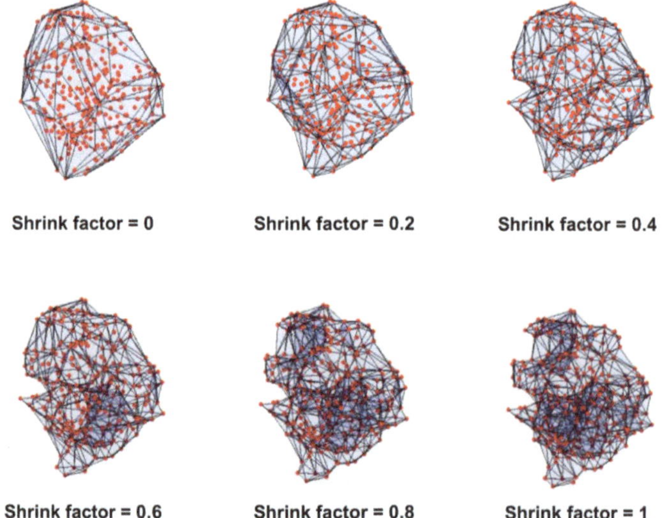

Figure 2. Dependence of the generated external boundary (external protein surface) on the value of the shrink factor. Infrared fluorescent protein (PDB: 5AJG) is reported as an example, with shrink factors equal to 0, 0.2, 0.4, 0.6, 0.8 and 1. Red points represent the nodes of the network (C^{α} atoms), while the external surface is represented by Delaunay triangles (in light blue), which connect the nodes in the external boundary.

For each of these generated surfaces, the random forces were applied only to the nodes lying on the external boundary. The resulting network displacements were then evaluated from Equations (4)–(7). It follows that the correlation of the average displacements (from 10,000 samples) due to the application of the external forces, i.e., $\rho_{FR,EXT}$, also depends on the adopted shrink factor. The shrink factor that leads to the maximum value of $\rho_{FR,EXT}$ for each protein is then selected as the optimal one. Figure 3 reports the statistical distribution of the optimal shrink factors obtained for the 919 single-chain proteins. As can be observed, the optimal shrink factor assumes almost all values, meaning that it is strongly protein-specific. Nevertheless, a slight preference towards shrink factors equal to 1.0 is observed.

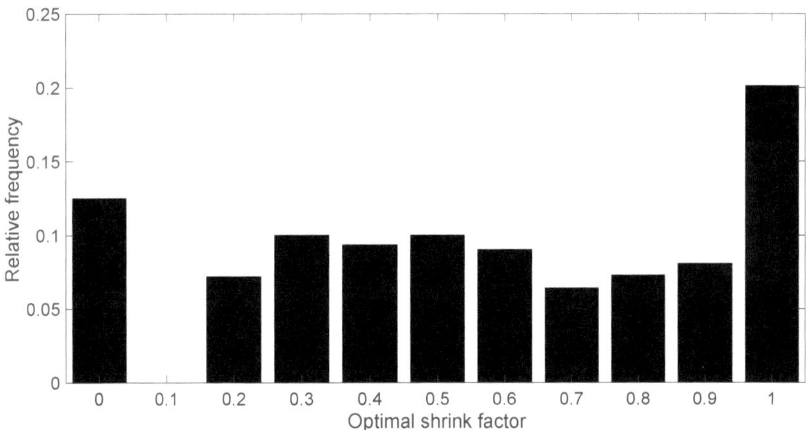

Figure 3. Statistical distribution of the optimal shrink factor, based on a comparison between experimental B-factors and average displacements due to the application of random forces on the exterior protein surface.

In Figure 4 we show results for the example of the infrared fluorescent protein (PDB: 5AJG), where the correlation coefficients ρ_{FL}, $\rho_{FR,ALL}$ and $\rho_{FR,EXT}$, are shown depending on the shrink factor for the external surface representation. From these calculations, we obtain a correlation between the B-factors and the traditional mode-based fluctuations ρ_{FL} equal to 0.56, a correlation with the displacements resulting from the application of forces to the entire structure $\rho_{FR,ALL}$ equal to 0.62, and a correlation derived from perturbations only on the external surface $\rho_{FR,EXT}$ which varies with the shrink factor and reaches a maximum value of 0.62 for an optimal shrink factor of 1.0. As can be seen from the results, in this case, applying random forces on the protein network enhances the correlation with the experimental B-factors of 6% (0.62 vs. 0.56) compared with the usual mode-based fluctuations. This result points out the cohesive nature of the protein structure, showing that the point of application of forces does not matter much, with the result of applying forces in all possible directions on the surface yielding nearly the same result as applying them in all directions throughout the structure when the surface representation is detailed enough.

Figures S1–S5 in the Supplementary Material report similar results, obtained by adopting different values of p for the decay exponent of the ENM spring constants ($p = 1, 2, 4, 6$ and 12). As can be seen there, similar conclusions can be drawn for these cases. Note that, for this protein, higher values of p, e.g., $p = 4$ and 6, lead to a greater enhancement in the correlation with experimental B-factors when the ENM forces are applied, rather than just looking at the intrinsic dynamical fluctuations.

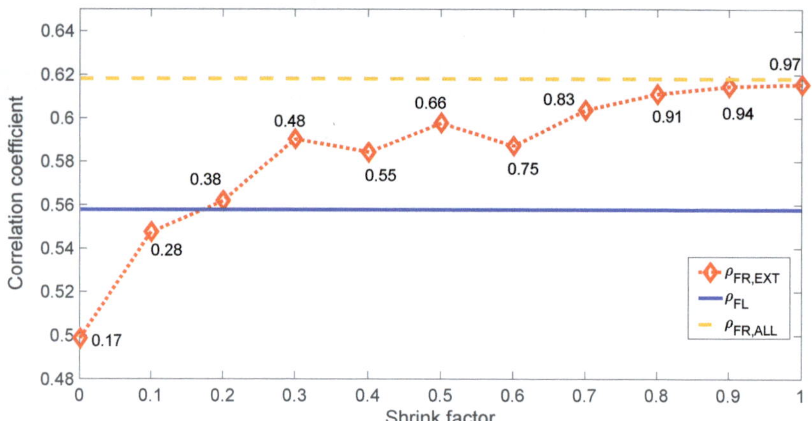

Figure 4. Experimental B-factors vs. mode-based fluctuations and average displacements due to random forces for the infrared fluorescent protein (PDB: 5AJG). The solid blue line refers to ρ_{FL}, the dashed orange line to $\rho_{FR,ALL}$ and the dotted red line to $\rho_{FR,EXT}$. The correlation arising from the application of forces only on the external surface, i.e., $\rho_{FR,EXT}$, depends on the selected shrink factor, which is in the range 0–1. For each shrink factor, the values reported close to the marker represent the fraction of external nodes, out of the total 301 nodes of the network. This shows that the best result is the structure representation with the greatest level of detail, and interestingly the most detailed structure with forces applied to the surface only leads to nearly the same result as the application of forces throughout the structure.

3.2. Incorporating Waters into the Computations

In this section, we show results obtained by also including the localized waters given in the crystal structure as part of the structure for defining the elastic network. There is some ongoing debate about whether or not these bound waters should be considered as an actual part of the structure. Each high-quality protein structure available in the PDB contains a substantial number of waters that were present in the crystal formed at low temperatures. The open question is whether these remain bound at higher temperatures. These molecules often typically form a network of hydrogen bonds with the side chains of polar amino acids on the protein surface and thus can appear to be quite stable. It follows that these waters may possibly cause some changes to the overall flexibility and dynamics of any given protein. Moreover, since we are interested in looking at the responses of each protein structure due to external perturbations, the inclusion of these external waters would be expected to affect the motions to some extent. Figure 5 shows a surface representation of the infrared fluorescent protein (PDB: 5AJG), with and without the addition of water molecules in Figure 5a,b, respectively. The protein structure is shown in light-gray, with the surface depiction highlighting the external surface and cavities. Water molecules available in the PDB are shown in Figure 5a as blue spheres. As can be seen from the comparison between the two figures, most of the crystallized water molecules are bound in concave parts of the protein surface, and thus act to smooth the structure [79]. This smoothing might restrict the flexibility of certain parts of the proteins that might cause problems for the mechanisms otherwise; this can be looked at as flowing liquid that represents protein motion, if a certain direction of the flow is restricted, it may change the overall flow of the water, therefore, to have an optimal flow path (specific functional protein motion), the restrictions (waters) are as important as the structure itself.

As mentioned in the previous section, the wpfANM is built as the usual pfANM with p equal to 3 (see Equation (2)), except that both the coordinates of the C^α atoms and the water molecules are now considered as a part of the whole structure. Based on the resulting wpfANM, the mode-based fluctuations are evaluated from Equation (3), whereas the average displacements resulting from the random perturbations are computed according to Equations (4)–(7). In the latter cases, as explained above, three types of force application are considered, as shown in Table 1. From the calculations, we then obtain four correlation coefficients to compare with the experimental B-factors, namely $\rho_{W,FL}$, $\rho_{W,FR,ALL}$, $\rho_{W,FR,EXT}$ and $\rho_{W,FR,WAT}$, with these being described in Table 1.

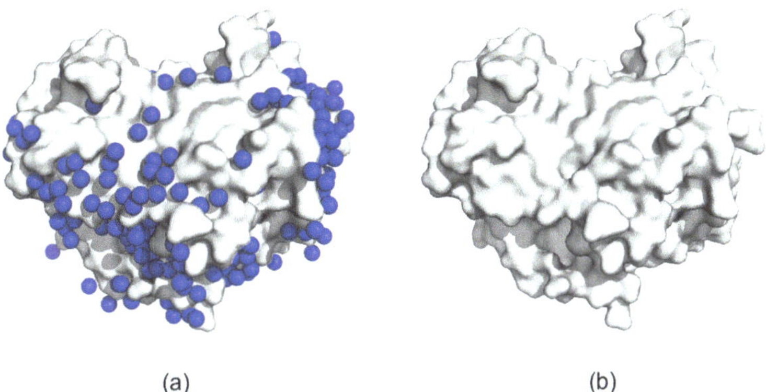

Figure 5. Surface representation of infrared fluorescent protein (PDB: 5AJG): (**a**) protein structure (light gray) + water molecules (blue spheres); (**b**) protein structure alone.

Figure 6 shows the correlation coefficients for the 919 proteins investigated. Figure 6a–c report the correlations ordered by increasing values of $\rho_{W,FL}$, whereas Figure 6d displays the statistical distribution for all four Pearson coefficients, whose median values and standard deviations are shown in the key. From the results reported in Figure 6, it follows that applying forces on the protein network slightly (<10%) enhances the prediction of the B-factors with respect to the traditional mode-based fluctuations. As a matter of fact, the median value of $\rho_{W,FL}$ was found to be 0.57 for the selected dataset, whereas the median values of $\rho_{W,FR,ALL}$, $\rho_{W,FR,EXT}$ and $\rho_{W,FR,WAT}$ were 0.60, 0.66, and 0.59, respectively. It is evident that applying random forces on the surface of the network (which now considers also the layer of water molecules) yields a significant 10% gain in the correlation with the experimental fluctuations, compared to the mode-based fluctuations.

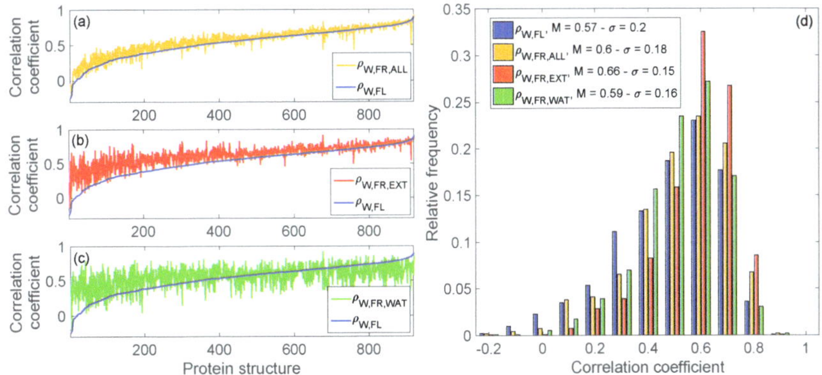

Figure 6. Experimental B-factors vs. mode-based fluctuations and average displacements due to random perturbations (wpfANM—water molecules included): (**a**–**c**) correlation coefficients for the 919 protein structures (blue curve for $\rho_{W,FL}$, orange curve for $\rho_{W,FR,ALL}$, red curve for $\rho_{W,FR,EXT}$ and green curve for $\rho_{W,FR,WAT}$); (**d**) statistical distribution of the correlation coefficients. The highest correlations are seen when perturbations are applied on the surface.

Also in the case of the wpfANM, the surface of the network is not unique but depends on the adopted shrink factor. The water molecules contained in the network play a major role in the definition of the surface since they are mostly placed on the exterior of the structure (see Figure 5). As an example, Figure 7 shows the different surfaces generated for the infrared fluorescent protein (PDB: 5AJG, with water molecules included) with shrink factors equal to 0, 0.2, 0.4, 0.6, 0.8 and 1. As can be seen by comparing Figure 7 to Figure 2, other than the selected value of the shrink factor, the shape of the external surface also depends on the presence of the water molecules within the network. Similarly to what was shown in the previous section, Figure 8 shows the optimal shrink factors (with the best correlation with B-factors) obtained for the 919 water-enriched protein structures. Again, the distribution of the optimal shrink factor is rather uniform, although in this case a slight bias towards the convex hulls surfaces appears, i.e., with a shrink factor equal to 0. This probably reflects a preference for smoother structures when water molecules are added (see Figure 5).

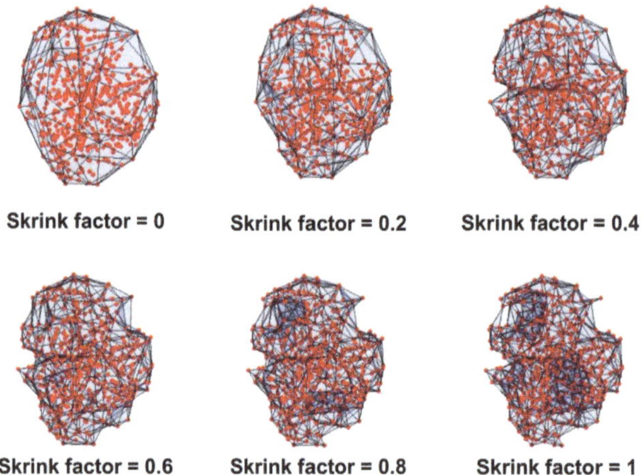

Figure 7. Dependence of the generated external boundary (external protein surface) on the value of the shrink factor. Infrared fluorescent protein (PDB: 5AJG), with water molecules included from the PDB structure file, is reported as an example, with shrink factors equal to 0, 0.2, 0.4, 0.6, 0.8 and 1. Red points represent the nodes of the network (C^α atoms + water molecules), while the external surface is represented by Delaunay triangles (in light blue), which connect the nodes in the external boundary.

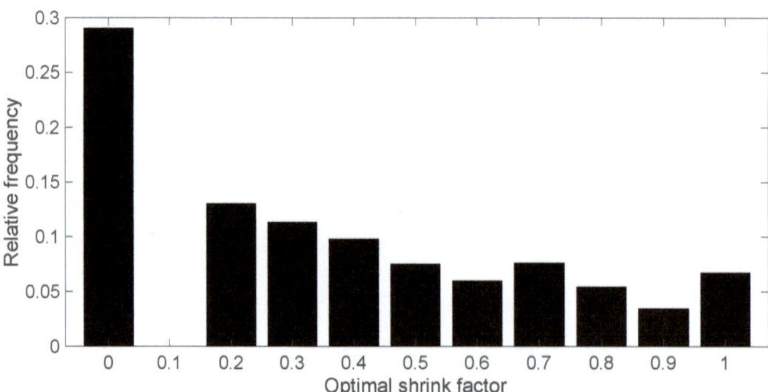

Figure 8. Statistical distribution of the optimal shrink factor, resulting from the comparison between experimental B-factors and average displacements due to the application of random forces on the external protein surface, with PDB water molecules included.

Figure 9 shows the correlation coefficients of the wpfANM of the infrared fluorescent protein (PDB: 5AJG), and how it depends on the adopted shrink factor for the surface. From the calculations, we obtain a correlation with the mode-based fluctuations $\rho_{W,FL}$ equal to 0.84, a correlation with the displacements from the application of forces to the entire structure $\rho_{W,FR,ALL}$ equal to 0.85, a correlation with the displacements from the application of forces only to the water molecules $\rho_{W,FR,WAT}$ equal to 0.63, and a correlation with the displacements due to the perturbations only on the surface $\rho_{W,FR,EXT}$ that varies with the shrink factor and reaches a maximum of 0.83 for the optimal shrink factor of 0.3. In this case, it is remarkable that by applying random perturbations on only 31% of the nodes (corresponding to a shrink factor of 0.3), we obtain a high correlation with the experimental data (0.83 Pearson coefficient). It should be noted that this correlation is found by comparing the experimental B-factors of all protein residues (both on the surface and within the core) to the computed average displacements due to the application of perturbations only on the external part of the elastic network. Thus, it follows that even perturbing a small portion of the protein surface (31%) allows us to predict fairly accurately the fluctuations of the entire protein. This result has its origin in the strong coupling throughout the elastic network model: since the ENM is a highly cooperative model, the perturbation of only a small part of the structure can indeed generate fluctuations over the entire protein. This arises from the specific features of the three-dimensional protein structure and all of the internal connections, which are both built into the ENM.

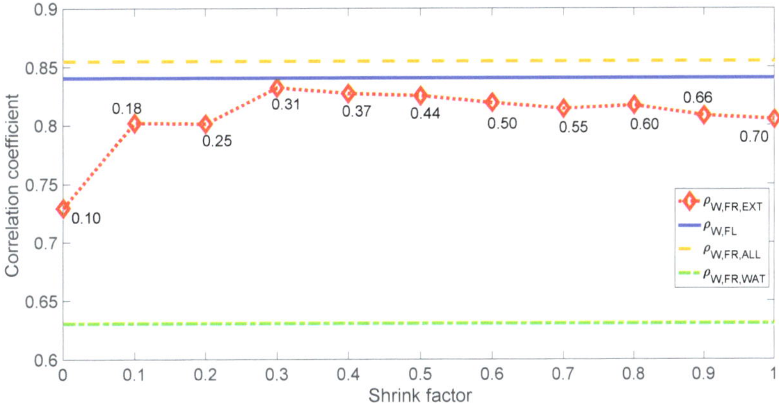

Figure 9. Experimental B-factors vs. mode-based fluctuations and average displacements due to random forces for the infrared fluorescent protein (PDB: 5AJG), with PDB water molecules included. The continuous blue line refers to $\rho_{W,FL}$, the dashed orange line to $\rho_{W,FR,ALL}$, the dotted red line to $\rho_{W,FR,EXT}$, and the dashed-dotted green line to $\rho_{W,FR,WAT}$. The correlations from the application of forces only on the protein's external surface, i.e., $\rho_{W,FR,EXT}$, depends on the selected shrink factor, in the range 0–1. For each shrink factor, the values reported close to the marker represent the fraction of external nodes out of the total of 533 nodes (301 C^{α} atoms + 252 water molecules) of the network.

Figures S6–S10 in the Supplementary Material show similar results as those reported in Figure 9, but for different exponents p in the water-augmented ENM, namely, $p = 1, 2, 4, 6$ and 12. Despite some differences in the numerical values of the correlation coefficients, similar conclusions are reached as for those in Figure 9.

3.3. Comparison between pfANM and wpfANM Results

In the previous sections, it has been shown that perturbing the protein structure with random forces, either on the entire structure or on the surface, generally leads to a fairly accurate prediction of the protein fluctuations and flexibility. In a large number of cases, it has also been found that the agreement with the experimental B-factors was improved with respect to considering the traditional fluctuations of the usual elastic network. As an example, the results shown in the previous sections for the infrared fluorescent protein (PDB: 5AJG) are reported together in Figure 10 in terms of correlation coefficients with the experimental B-factors. Several observations can be made regarding this figure. First, it is evident that the traditionally employed internal protein fluctuations provide the lowest correlation with the experimental data, with a correlation of about 55% (ρ_{FL}). Conversely, applying random perturbations on the same elastic network leads to a 5% gain in the correlations with the B-factors ($\rho_{FR,ALL}$ and $\rho_{FR,EXT}$). Furthermore, adding the PDB water molecules to the elastic network further improves the correlation with the experimental data. In this case, considering the internal protein fluctuations of the water-enriched elastic network or applying random forces to it yields correlation coefficients of about 85% ($\rho_{W,FL}$, $\rho_{W,FR,ALL}$ and $\rho_{W,FR,EXT}$). Also, applying random forces only on these water molecules, i.e., not perturbing the protein molecule directly but only the water molecules attached to the network (see Figure 5a), leads to a correlation of about 60% ($\rho_{W,FR,WAT}$), which is still higher than the correlation obtained with the traditional mode-based internal protein fluctuations of the ENM (ρ_{FL} = 55%).

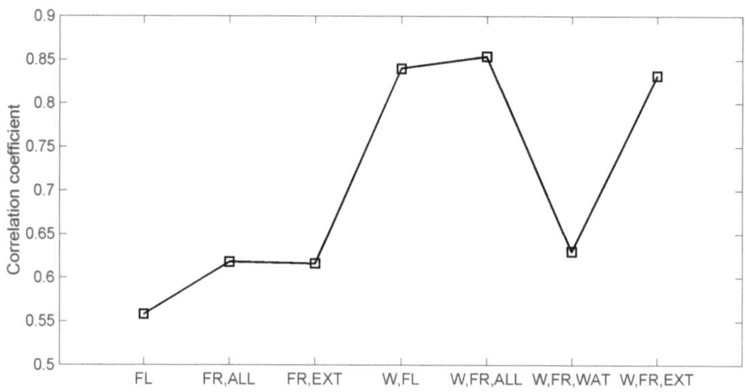

Figure 10. Infrared fluorescent protein (PDB: 5AJG): correlation coefficients obtained from the seven types of analyses, as reported in Table 1 and in the previous sections.

Figures S11–S15 in the Supplementary Material show similar outcomes, but are obtained by changing the exponent p of the ENM. As can be seen, similar conclusions are drawn. In all cases, it is found that perturbing the protein with random forces provides an enhancement of the correlation with experimental B-factors, rather than considering the traditional ENM with only intra-protein interactions. Moreover, the inclusion of waters in the ENM leads to further improvements in the correlation, with some gains in the Pearson coefficient being as high as 30–35%.

The example shown in Figure 10 obviously refers to one single case, but these results were found for a decent amount of protein structures. For other protein structures, the addition of water molecules to the protein network led to results which were quite similar to the ones obtained in the classical way, i.e., calculating the mode-based fluctuations of the traditional no-water elastic network.

Figure 11a reports the median values and standard deviations of the seven correlation coefficients obtained for the dataset of 919 single-chain protein structures, as reported in the keys of Figures 1c and 6d. As can be observed, the median values lie in the range 0.60–0.65, and present a similar standard deviation (15–20%). However, the distribution with the highest median value and the lowest standard deviation was found for the analysis involving the application of random perturbation on the external surface of the wpfANM, i.e., $\rho_{W,FR,EXT}$. A direct comparison between the statistical distribution of $\rho_{W,FR,EXT}$ and the one related to the traditional mode-based internal protein fluctuations of the no-water pfANM, i.e., $\rho_{W,FL}$, is reported in Figure 11b. From the direct comparison of the two distributions, it is clear that applying random forces on the surface of the wpfANM leads to an overall, yet slight, improvement of the correlation with the experimental data.

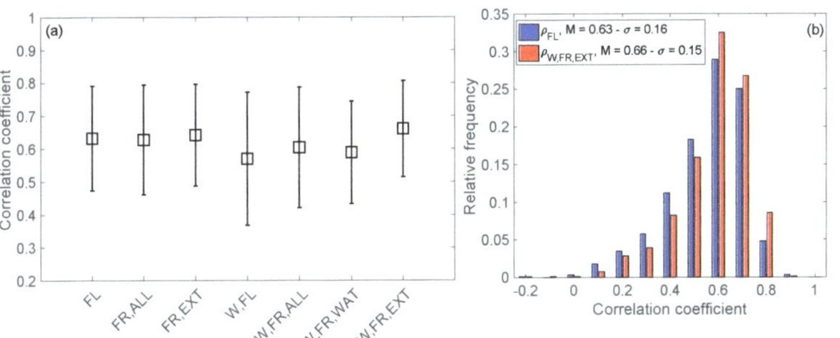

Figure 11. (**a**) Correlation coefficients obtained from the seven types of analyses (median values and standard deviations of the statistical distributions) for the 919 single-chain protein structures; (**b**) direct comparison between the statistical distributions of ρ_{FL} and $\rho_{W,FR,EXT}$.

Based on the correlation coefficients obtained for our entire dataset of protein structures, the analysis among the seven performed ones (Table 1) that gave the highest correlations with the experimental data was noted. Figure 12a shows the relative number of such occurrences for each type of analysis. It was found that, out of the 919 investigated protein structures, the mode-based fluctuations of the pfANM provided the highest correlation coefficient in 107 cases (11.6%), the application of random forces on the entire pfANM in 74 cases (8.1%), the application of forces on the external surface of the pfANM for 172 cases (18.7%), the mode-based fluctuations of the wpfANM in 60 cases (6.5%), the application of forces on the entire wpfANM in 70 cases (7.62%), the application of forces only on the water molecules of the wpfANM in 93 cases (10.1%), while the application of forces on the external surface of the wpfANM were provided in 343 cases (37.3%). As can be seen and has already been discussed concerning Figure 11, the application of random forces on the surface of the water-enriched protein network is statistically the best performing, although not the only one, with regards to better predicting protein fluctuations and flexibility in terms of experimental B-factors. Moreover, looking cumulatively at the analyses *FR,EXT* and *W,FR,EXT*, in 515 cases (56.0%), the application of random forces on the external surface of the protein network yields the highest correlation coefficients, thus confirming that perturbing the external protein surface can induce a response in good agreement with the experimental B-factors.

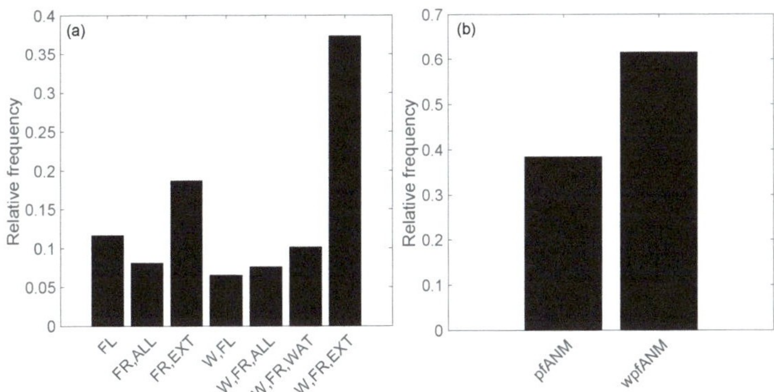

Figure 12. (**a**) Relative frequency of the highest correlation coefficient for each type of analysis for the dataset of 919 single-chain protein structures; (**b**) relative frequency of the highest correlation coefficient for the pfANM or the wpfANM.

Figure 12b shows the occurrence of the highest correlation coefficients for the two models, i.e., the pfANM vs. wpfANM. From the outcomes, it was obtained that in 353 cases (38.41%) the highest correlation with the B-factors was obtained with the pfANM, whereas in the remaining 566 cases (61.59%) the wpfANM was allowed to reach the highest correlation with the experimental data. This confirms that considering the PDB water molecules might actually enhance the prediction of the protein fluctuations and therefore the numerical evaluation of the experimental B-factors.

4. Conclusions

Research carried out in the last decades has shown that protein fluctuations and flexibility, as measured by the experimental B-factors, mainly arise from the internal protein motions and inherent dynamics. The dynamics are known to originate from the tertiary structure, as recognized within the fundamental sequence-structure-dynamics-function paradigm of protein action. Therefore, it should be more appropriate to say that protein fluctuations and flexibility arise from the protein structure and can be mediated by its dynamics. As a matter of fact, in a recent work [62], we have shown that the overall protein flexibility, as measured by the experimental B-factors, can also be retrieved by applying pairwise static forces to the protein ENM and measuring the amount of compliance against these external pulling forces.

In this paper we have proposed an additional viewpoint as regards protein fluctuations and flexibility. We applied random static forces throughout the protein elastic network and evaluated the response of the network via the computation of average nodal displacements. From the comparison of these average displacements against the experimental B-factors, we found that the protein flexibility, and therefore its fluctuations, can indeed be elucidated with such a procedure. Also, we found that if these perturbations are applied on the protein surface, and if crystallized water molecules are also inserted into the model, higher correlations with experimental data can often be found. This suggests that protein fluctuations can also be seen as (fully or partly) the response of the protein structure to external forces, which might be induced by the continuous collisions of the solvent and other solute molecules around the protein structure.

It is important to mention that the goal of the analysis presented here was to propose a new methodology, and therefore a new perspective, to understand protein fluctuations. However, no additional work has been carried out yet as regards the optimization of the elastic model upon which the random perturbations are applied. This might eventually improve the correlation with the experimental data and is our next goal. As described in the text, the pfANM has been used for the standard ENM, whereas a wpfANM has been generated in order to account for the presence of crystallized water, where water molecules have simply been added as additional nodes to the network. We plan to optimize the elastic model by including different spring constants for the C^α-C^α connections, C^α-water connections and water-water connections, which should simulate more realistically the different atomic interactions (residue-residue, residue-water, water-water).

Attention has been paid to the external surface of the network. Thus, we are also planning to use a different version of ENM, where the contact topology is not generated by using the traditional cut-off limit, but using alpha-shapes associated with Delaunay tessellations. A recent work from Koehl et al. [36] showed that such a procedure is able to generate elastic models with enhanced agreement with experimental data. Applying external forces on such optimized models, and adding water molecules as well, might enhance the correlation with experimental B-factors, allowing for the better explanation of fluctuations, and therefore the way a protein moves and functions.

Supplementary Materials: The following are available online at https://www.mdpi.com/article/10.3390/app12052344/s1, Figures S1–S5: Experimental B-factors vs. mode-based fluctuations and average displacements due to random forces for the infrared fluorescent protein (PDB: 5AJG), with $p = 1, 2, 4, 6$ and 12 for the decay of spring constants in the ENM, Figures S6–S10: Experimental B-factors vs. mode-based fluctuations and average displacements due to random forces for the infrared fluorescent protein (PDB: 5AJG), with PDB water molecules included, with $p = 1, 2, 4, 6$ and 12 for the decay of spring constants in the ENM, Figures S11–S15: Infrared fluorescent protein (PDB: 5AJG): correlation coefficients obtained from the seven types of analyses, as reported in Table 1 and in the previous sections. Case with $p = 1, 2, 4, 6$ and 12 for the decay of spring constants in the ENM.

Author Contributions: Conceptualization, D.S., P.M.K. and R.L.J.; methodology, D.S. and P.M.K.; software, D.S.; validation, D.S., P.M.K. and R.L.J.; formal analysis, D.S. and P.M.K.; investigation, D.S. and P.M.K.; resources, R.L.J.; data curation, D.S.; writing—original draft preparation, D.S.; writing— review and editing, D.S., P.M.K. and R.L.J.; visualization, D.S., P.M.K. and R.L.J.; supervision, R.L.J.; project administration, R.L.J.; funding acquisition, R.L.J. All authors have read and agreed to the published version of the manuscript.

Funding: We gratefully acknowledge the support from NIH grant R01GM127701.

Institutional Review Board Statement: Not applicable.

Informed Consent Statement: Not applicable.

Data Availability Statement: Additional data is available upon request from the authors.

Acknowledgments: We thank Research IT @Iowa State University for helping with some aspects of the computing.

Conflicts of Interest: The authors declare that they have no conflict of interest.

References

1. Debye, P. Interferenz von Röntgenstrahlen und Wärmebewegung. *Ann. Phys.* **1913**, *348*, 49–92. [CrossRef]
2. Trueblood, K.N.; Bürgi, H.-B.; Burzlaff, H.; Dunitz, J.D.; Gramaccioli, C.M.; Schulz, H.H.; Shmueli, U.; Abrahams, S.C. Atomic Dispacement Parameter Nomenclature. Report of a Subcommittee on Atomic Displacement Parameter Nomenclature. *Acta Crystallogr. Sect. A* **1996**, *52*, 770–781. [CrossRef]
3. Sherwood, D.; Cooper, J. *Crystals, X-rays and Proteins: Comprehensive Protein Crystallography*; OUP Oxford: Oxford, UK, 2010.
4. Na, H.; Hinsen, K.; Song, G. The Amounts of Thermal Vibrations and Static Disorder in Protein X-ray Crystallographic B-factors. *Proteins Struct. Funct. Bioinform.* **2021**, *89*, 1442–1457. [CrossRef] [PubMed]
5. Karplus, P.A.; Schulz, G.E. Prediction of chain flexibility in proteins—A tool for the selection of peptide antigens. *Naturwissenschaften* **1985**, *72*, 212–213. [CrossRef]
6. Schlessinger, A.; Rost, B. Protein flexibility and rigidity predicted from sequence. *Proteins Struct. Funct. Bioinform.* **2005**, *61*, 115–126. [CrossRef]
7. Yuan, Z.; Zhao, J.; Wang, Z.-X. Flexibility analysis of enzyme active sites by crystallographic temperature factors. *Protein Eng. Des. Sel.* **2003**, *16*, 109–114. [CrossRef]
8. Radivojac, P.; Obtadovic, Z.; Smith, D.K.; Zhu, G.; Vucetic, S.; Brown, C.J.; David Lawson, J.; Keith Dunker, A. Protein flexibility and intrinsic disorder. *Protein Sci.* **2004**, *13*, 71–80. [CrossRef]
9. Kuczera, K.; Kuriyan, J.; Karplus, M. Temperature dependence of the structure and dynamics of myoglobin. A simulation approach. *J. Mol. Biol.* **1990**, *213*, 351–373. [CrossRef]
10. Teilum, K.; Olsen, J.G.; Kragelund, B.B. Functional aspects of protein flexibility. *Cell. Mol. Life Sci.* **2009**, *66*, 2231. [CrossRef]
11. Huber, R.; Bennett, W.S., Jr. Functional significance of flexibility in proteins. *Biopolymers* **1983**, *22*, 261–279. [CrossRef]
12. Bahar, I.; Jernigan, R.L.; Dill, K.A. *Protein Actions: Principles & Modeling*; Garland Science: New York, NY, USA, 2017.
13. Karplus, M.; Kuriyan, J. Molecular dynamics and protein function. *Proc. Natl. Acad. Sci. USA* **2005**, *102*, 6679–6685. [CrossRef] [PubMed]
14. McCammon, J.A.; Gelin, B.R.; Karplus, M. Dynamics of folded proteins. *Nature* **1977**, *267*, 585–590. [CrossRef] [PubMed]
15. Hospital, A.; Goñi, J.R.; Orozco, M.; Gelpí, J.L. Molecular dynamics simulations: Advances and applications. *Adv. Appl. Bioinform. Chem.* **2015**, *8*, 37–47. [PubMed]
16. Meinhold, L.; Smith, J.C. Fluctuations and Correlations in Crystalline Protein Dynamics: A Simulation Analysis of Staphylococcal Nuclease. *Biophys. J.* **2005**, *88*, 2554–2563. [CrossRef]
17. Pang, Y.-P. Use of multiple picosecond high-mass molecular dynamics simulations to predict crystallographic B-factors of folded globular proteins. *Heliyon* **2016**, *2*, e00161. [CrossRef]
18. Go, N.; Noguti, T.; Nishikawa, T. Dynamics of a small globular protein in terms of low-frequency vibrational modes. *Proc. Natl. Acad. Sci. USA* **1983**, *80*, 3696–3700. [CrossRef]
19. Brooks, B.; Karplus, M. Harmonic dynamics of proteins: Normal modes and fluctuations in bovine pancreatic trypsin inhibitor. *Proc. Natl. Acad. Sci. USA* **1983**, *80*, 6571–6575. [CrossRef]
20. Levitt, M.; Sander, C.; Stern, P.S. Protein normal-mode dynamics: Trypsin inhibitor, crambin, ribonuclease and lysozyme. *J. Mol. Biol.* **1985**, *181*, 423–447. [CrossRef]
21. Ben-Avraham, D. Vibrational normal-mode spectrum of globular proteins. *Phys. Rev. B* **1993**, *47*, 14559–14560. [CrossRef]

22. Dykeman, E.C.; Sankey, O.F. Normal mode analysis and applications in biological physics. *J. Phys. Condens. Matter* **2010**, *22*, 423202. [CrossRef]
23. Bahar, I.; Cui, Q. *Normal Mode Analysis: Theory and Applications to Biological and Chemical Systems*; Chapman & Hall: London, UK, 2006.
24. Dehouck, Y.; Bastolla, U. Why are large conformational changes well described by harmonic normal modes? *Biophys. J.* **2021**, *120*, 5343–5354. [CrossRef] [PubMed]
25. Tirion, M.M. Large amplitude elastic motions in proteins from a single-parameter, atomic analysis. *Phys. Rev. Lett.* **1996**, *77*, 1905–1908. [CrossRef] [PubMed]
26. Haliloglu, T.; Bahar, I.; Erman, B. Gaussian dynamics of folded proteins. *Phys. Rev. Lett.* **1997**, *79*, 3090–3093. [CrossRef]
27. Bahar, I.; Atilgan, A.R.; Erman, B. Direct evaluation of thermal fluctuations in proteins using a single-parameter harmonic potential. *Fold. Des.* **1997**, *2*, 173–181. [CrossRef]
28. Rader, A.J.; Chennubhotla, C.; Yang, L.-W.; Bahar, I. The Gaussian Network Model: Theory and Applications. In *Normal Mode Analysis: Theory and Applications to Biological and Chemical Systems*; Cui, Q., Bahar, I., Eds.; Chapman & Hall: London, UK, 2006; pp. 41–64.
29. Micheletti, C.; Carloni, P.; Maritan, A. Accurate and Efficient Description of Protein Vibrational Dynamics: Comparing Molecular Dynamics and Gaussian Models. *Proteins Struct. Funct. Genet.* **2004**, *55*, 635–645. [CrossRef] [PubMed]
30. Bahar, I.; Erman, B.; Jernigan, R.L.; Atilgan, A.R.; Covell, D.G. Collective motions in HIV-1 reverse transcriptase: Examination of flexibility and enzyme function. *J. Mol. Biol.* **1999**, *285*, 1023–1037. [CrossRef] [PubMed]
31. Bahar, I.; Jernigan, R.L. Cooperative fluctuations and subunit communication in tryptophan synthase. *Biochemistry* **1999**, *38*, 3478–3490. [CrossRef]
32. Atilgan, A.R.; Durell, S.R.; Jernigan, R.L.; Demirel, M.C.; Keskin, O.; Bahar, I. Anisotropy of fluctuation dynamics of proteins with an elastic network model. *Biophys. J.* **2001**, *80*, 505–515. [CrossRef]
33. Yang, L.; Song, G.; Jernigan, R.L. Protein elastic network models and the ranges of cooperativity. *Proc. Natl. Acad. Sci. USA* **2009**, *106*, 12347–12352. [CrossRef]
34. Eyal, E.; Yang, L.W.; Bahar, I. Anisotropic network model: Systematic evaluation and a new web interface. *Bioinformatics* **2006**, *22*, 2619–2627. [CrossRef]
35. Kim, M.H.; Lee, B.H.; Kim, M.K. Robust elastic network model: A general modeling for precise understanding of protein dynamics. *J. Struct. Biol.* **2015**, *190*, 338–347. [CrossRef] [PubMed]
36. Koehl, P.; Orland, H.; Delarue, M. Parameterizing elastic network models to capture the dynamics of proteins. *J. Comput. Chem.* **2021**, *42*, 1643–1661. [CrossRef] [PubMed]
37. Orellana, L.; Rueda, M.; Ferrer-Costa, C.; Lopez-Blanco, J.R.; Chacón, P.; Orozco, M. Approaching elastic network models to molecular dynamics flexibility. *J. Chem. Theory Comput.* **2010**, *6*, 2910–2923. [CrossRef] [PubMed]
38. Tama, F.; Sanejouand, Y.H. Conformational change of proteins arising from normal mode calculations. *Protein Eng.* **2001**, *14*, 1–6. [CrossRef] [PubMed]
39. Yang, L.; Song, G.; Jernigan, R.L. How well can we understand large-scale protein motions using normal modes of elastic network models? *Biophys. J.* **2007**, *93*, 920–929. [CrossRef]
40. Petrone, P.; Pande, V.S. Can conformational change be described by only a few normal modes? *Biophys. J.* **2006**, *90*, 1583–1593. [CrossRef] [PubMed]
41. Mahajan, S.; Sanejouand, Y.H. On the relationship between low-frequency normal modes and the large-scale conformational changes of proteins. *Arch. Biochem. Biophys.* **2015**, *567*, 59–65. [CrossRef]
42. Sanejouand, Y.-H. Normal-mode driven exploration of protein domain motions. *J. Comput. Chem.* **2021**, *42*, 2250–2257. [CrossRef]
43. Mahajan, S.; Sanejouand, Y.H. Jumping between protein conformers using normal modes. *J. Comput. Chem.* **2017**, *38*, 1622–1630. [CrossRef]
44. Zheng, W.; Doniach, S. A comparative study of motor-protein motions by using a simple elastic-network model. *Proc. Natl. Acad. Sci. USA* **2003**, *100*, 13253–13258. [CrossRef]
45. Zheng, W.; Brooks, B.R. Normal-modes-based prediction of protein conformational changes guided by distance constraints. *Biophys. J.* **2005**, *88*, 3109–3117. [CrossRef] [PubMed]
46. Dobbins, S.E.; Lesk, V.I.; Sternberg, M.J.E. Insights into protein flexibility: The relationship between normal modes and conformational change upon protein-protein docking. *Proc. Natl. Acad. Sci. USA* **2008**, *105*, 10390–10395. [CrossRef]
47. Tobi, D.; Bahar, I. Structural changes involved in protein binding correlate with intrinsic motions of proteins in the unbound state. *Proc. Natl. Acad. Sci. USA* **2005**, *102*, 18908–18913. [CrossRef] [PubMed]
48. Khade, P.M.; Scaramozzino, D.; Kumar, A.; Lacidogna, G.; Carpinteri, A.; Jernigan, R.L. hdANM: A new comprehensive dynamics model for protein hinges. *Biophys. J.* **2021**, *120*, 4955–4965. [CrossRef] [PubMed]
49. Kim, M.K.; Chirikjian, G.S.; Jernigan, R.L. Elastic models of conformational transitions in macromolecules. *J. Mol. Graph. Model.* **2002**, *21*, 151–160. [CrossRef]
50. Kim, M.K.; Jernigan, R.L.; Chirikjian, G.S. Efficient generation of feasible pathways for protein conformational transitions. *Biophys. J.* **2002**, *83*, 1620–1630. [CrossRef]
51. Kim, M.K.; Jernigan, R.L.; Chirikjian, G.S. Rigid-cluster models of conformational transitions in macromolecular machines and assemblies. *Biophys. J.* **2005**, *89*, 43–55. [CrossRef]

52. Maragakis, P.; Karplus, M. Large amplitude conformational change in proteins explored with a plastic network model: Adenylate kinase. *J. Mol. Biol.* **2005**, *352*, 807–822. [CrossRef]

53. Eom, K. Conformational Changes of Protein Analyzed Based on Structural Perturbation Method. *Multiscale Sci. Eng.* **2021**, *3*, 62–66. [CrossRef]

54. Orellana, L.; Yoluk, O.; Carrillo, O.; Orozco, M.; Lindahl, E. Prediction and validation of protein intermediate states from structurally rich ensembles and coarse-grained simulations. *Nat. Commun.* **2016**, *7*, 12575. [CrossRef]

55. Ikeguchi, M.; Ueno, J.; Sato, M.; Kidera, A. Protein structural change upon ligand binding: Linear response theory. *Phys. Rev. Lett.* **2005**, *94*, 078102. [CrossRef]

56. Atilgan, C.; Atilgan, A.R. Perturbation-response scanning reveals ligand entry-exit mechanisms of ferric binding protein. *PLoS Comput. Biol.* **2009**, *5*, e1000544. [CrossRef]

57. Atilgan, C.; Gerek, Z.N.; Ozkan, S.B.; Atilgan, A.R. Manipulation of conformational change in proteins by single-residue perturbations. *Biophys. J.* **2010**, *99*, 933–943. [CrossRef] [PubMed]

58. Gerek, Z.N.; Ozkan, S.B. Change in allosteric network affects binding affinities of PDZ domains: Analysis through perturbation response scanning. *PLoS Comput. Biol.* **2011**, *7*, e1002154. [CrossRef] [PubMed]

59. Liu, J.; Sankar, K.; Wang, Y.; Jia, K.; Jernigan, R.L. Directional Force Originating from ATP Hydrolysis Drives the GroEL Conformational Change. *Biophys. J.* **2017**, *112*, 1561–1570. [CrossRef] [PubMed]

60. Scaramozzino, D.; Piana, G.; Lacidogna, G.; Carpinteri, A. Low-Frequency Harmonic Perturbations Drive Protein Conformational Changes. *Int. J. Mol. Sci.* **2021**, *22*, 10501. [CrossRef]

61. Eyal, E.; Bahar, I. Toward a molecular understanding of the anisotropic response of proteins to external forces: Insights from elastic network models. *Biophys. J.* **2008**, *94*, 3424–3435. [CrossRef]

62. Scaramozzino, D.; Khade, P.M.; Jernigan, R.L.; Lacidogna, G.; Carpinteri, A. Structural Compliance: A New Metric for Protein Flexibility. *Proteins Struct. Funct. Bioinform.* **2020**, *88*, 1482–1492. [CrossRef]

63. Sen, T.Z.; Feng, Y.; Garcia, J.V.; Kloczkowski, A.; Jernigan, R.L. The extent of cooperativity of protein motions observed with elastic network models is similar for atomic and coarser-grained models. *J. Chem. Theory Comput.* **2006**, *2*, 696–704. [CrossRef]

64. Frey, M. Water structure associated with proteins and its role in crystallization. *Acta Crystallogr. Sect. D Biol. Crystallogr.* **1994**, *50*, 663–666. [CrossRef]

65. Bhat, T.N.; Bentley, G.A.; Boulot, G.; Greene, M.I.; Tello, D.; Dall'Acqua, W.; Souchon, H.; Schwarz, F.P.; Mariuzza, R.A.; Poljak, R.J. Bound water molecules and conformational stabilization help mediate an antigen-antibody association. *Proc. Natl. Acad. Sci. USA* **1994**, *91*, 1089–1093. [CrossRef] [PubMed]

66. Hayward, S.; Kitao, A.; Hirata, F.; Go, N. Effect of solvent on collective motions in globular protein. *J. Mol. Biol.* **1993**, *234*, 1207–1217. [CrossRef] [PubMed]

67. Chandler, D. Interfaces and the driving force of hydrophobic assembly. *Nature* **2005**, *437*, 640–647. [CrossRef] [PubMed]

68. Nakasako, M. Large-scale networks of hydration water molecules around bovine β-trypsin revealed by cryogenic X-ray crystal structure analysis. *J. Mol. Biol.* **1999**, *289*, 547–564. [CrossRef] [PubMed]

69. Lins, L.; Thomas, A.; Brasseur, R. Analysis of accessible surface of residues in proteins. *Protein Sci.* **2003**, *12*, 1406–1417. [CrossRef]

70. Prabhu, N.; Sharp, K. Protein-solvent interactions. *Chem. Rev.* **2008**, *106*, 1616–1623. [CrossRef]

71. Brysbaert, G.; Blossey, R.; Lensink, M.F. The inclusion of water molecules in residue interaction networks identifies additional central residues. *Front. Mol. Biosci.* **2018**, *5*, 88. [CrossRef]

72. Horvath, I.; Jeszenoi, N.; Balint, M.; Paragi, G.; Hetenyi, C. A fragmenting protocol with explicit hydration for calculation of binding enthalpies of target-ligand complexes at a quantum mechanical level. *Int. J. Mol. Sci.* **2019**, *20*, 4384. [CrossRef]

73. Berman, H.M.; Westbrook, J.; Feng, Z.; Gilliland, G.; Bhat, T.N.; Weissig, H.; Shindyalov, I.N. The Protein Data Bank. *Nucleic Acids Res.* **2000**, *28*, 235–242. [CrossRef]

74. Eyal, E.; Lum, G.; Bahar, I. The anisotropic network model web server at 2015 (ANM 2.0). *Bioinformatics* **2015**, *31*, 1487–1489. [CrossRef]

75. Scaramozzino, D.; Lacidogna, G.; Piana, G.; Carpinteri, A. A finite-element-based coarse-grained model for global protein vibration. *Meccanica* **2019**, *54*, 1927–1940. [CrossRef]

76. Giordani, G.; Scaramozzino, D.; Iturrioz, I.; Lacidogna, G.; Carpinteri, A. Modal analysis of the lysozyme protein considering all-atom and coarse-grained finite element models. *Appl. Sci.* **2021**, *11*, 547. [CrossRef]

77. Khade, P.M.; Kumar, A.; Jernigan, R.L. Characterizing and Predicting Protein Hinges for Mechanistic Insight. *J. Mol. Biol.* **2020**, *432*, 508–522. [CrossRef] [PubMed]

78. Kurkcuoglu, O.; Jernigan, R.L.; Doruker, P. Mixed levels of coarse-graining of large proteins using elastic network model succeeds in extracting the slowest motions. *Polymer* **2004**, *45*, 649–657. [CrossRef]

79. Tsai, J.; Taylor, R.; Chothia, C.; Gerstein, M. The packing density in proteins: Standard radii and volumes. *J. Mol. Biol.* **1999**, *290*, 253–266. [CrossRef]

applied sciences

MDPI

Article

In Search of a Dynamical Vocabulary: A Pipeline to Construct a Basis of Shared Traits in Large-Scale Motions of Proteins

Thomas Tarenzi [1,2], Giovanni Mattiotti [1,2], Marta Rigoli [3] and Raffaello Potestio [1,2,*]

[1] Department of Physics, University of Trento, Via Sommarive 14, 38123 Trento, Italy; thomas.tarenzi@unitn.it (T.T.); giovanni.mattiotti@unitn.it (G.M.)

[2] INFN-TIFPA, Trento Institute for Fundamental Physics and Applications, Via Sommarive 14, 38123 Trento, Italy

[3] Centre for Integrative Biology (CIBIO), University of Trento, Via Sommarive 9, 38123 Trento, Italy; marta.rigoli@unitn.it

* Correspondence: raffaello.potestio@unitn.it; Tel.: +39-0461-282912

Abstract: The paradigmatic sequence–structure–dynamics–function relation in proteins is currently well established in the scientific community; in particular, a large effort has been made to probe the first connection, indeed providing convincing evidence of its strength and rationalizing it in a quantitative and general framework. In contrast, however, the role of dynamics as a link between structure and function has eluded a similarly clear-cut verification and description. In this work, we propose a pipeline aimed at building a basis for the quantitative characterization of the large-scale dynamics of a set of proteins, starting from the sole knowledge of their native structures. The method hinges on a dynamics-based clusterization, which allows a straightforward comparison with structural and functional protein classifications. The resulting basis set, obtained through the application to a group of related proteins, is shown to reproduce the salient large-scale dynamical features of the dataset. Most interestingly, the basis set is shown to encode the fluctuation patterns of homologous proteins not belonging to the initial dataset, thus highlighting the general applicability of the pipeline used to build it.

Keywords: protein dynamics; elastic network models; normal mode analysis

Citation: Tarenzi, T.; Mattiotti, G.; Rigoli, M.; Potestio, R. In Search of a Dynamical Vocabulary: A Pipeline to Construct a Basis of Shared Traits in Large-Scale Motions of Proteins. *Appl. Sci.* **2022**, *12*, 7157. https://doi.org/10.3390/app12147157

Academic Editors: Robert Jernigan and Domenico Scaramozzino

Received: 20 June 2022
Accepted: 13 July 2022
Published: 15 July 2022

Publisher's Note: MDPI stays neutral with regard to jurisdictional claims in published maps and institutional affiliations.

1. Introduction

The internal motions of proteins are intimately linked to protein function [1]. Such conformational movements span a wide range of spatial and temporal scales, going from local sidechain rotations and loop motions (ps to ns), to conformational transitions involving unfolding/refolding processes (ms to hours) [2]. In between these two extremes, internal large-scale protein fluctuations happening on timescales of the order of ns-μs [3] typically involve the collective movements of secondary structure elements; such fluctuations lead to a variety of potential conformational states, which might promote the exposure of specific binding sites [4,5] or facilitate the induced fit of the protein upon interaction with partner molecules [6,7]. It has been shown not only that this large-scale dynamics is essential for a protein to carry out its biological role [8], but also that a remarkable correlation exists between a protein's function and its specific dynamical signature [9], thus strengthening the view of dynamics as a link between a protein's structure and its specific function. This is particularly evident for the case of allosteric proteins, where the binding of a ligand conveys a signal that is propagated within the protein structure through a modulation of its internal dynamics, resulting in alternative conformational states and an altered protein function [10–12].

Several computational methods exist for the study of collective dynamics in proteins [13–15]; however, in order to develop a more general view of how dynamics bridges structure and function, it is necessary to build a datasetwise approach for the comparison of such large-scale dynamics among proteins sharing different degrees of sequence and structural similarity. Attempts in this direction have been performed in several works [16–21]. Maguid et al. [22] based their analysis on a dataset of pairs of homologous proteins; the comparison of vibrational backbone dynamics within each pair led to the remarkable observation of a correlation between dynamics and evolutionary conservation. Velázquez-Muriel et al. [23] performed a comparison between the protein flexibility shown by the structurally aligned members of a CATH superfamily [24] and the protein flexibility sampled by molecular dynamics simulation of a reference protein belonging to the same superfamily. Singular-value decomposition was used to capture the essential components of the two spaces, which show different size and complexity and are therefore suggested to be combined for a thorough exploration of protein deformations. Analyses of the distance in dynamics have also been performed in the case of structurally and functionally diverse sets of proteins; in this regard, Hensen et al. [9] introduced the notion of the "dynasome", namely an ensemble of observables computed from molecular dynamics (MD) simulations of a structurally heterogeneous protein dataset. The method highlights a striking correlation between the dynasome descriptors (which include 34 observables for each protein, ranging from the first five eigenvalues of the covariance matrix of C_α fluctuations to the average ruggedness of the energy landscape) and the proteins' functional classification. However, this approach relies on time-consuming MD simulations, which limits its applicability to large protein datasets. In addition, the large number and sophistication of the descriptors employed do not enable a straightforward recognition and visualization of the similarities in dynamics between proteins in terms of conformational movements.

To overcome these limitations, in this work, we set up and validate a novel pipeline for the identification of a basis set of conformational motions in an enzymatic family, representing a common vocabulary of their large-scale dynamics. To this aim, we investigated internal, collective protein dynamics in terms of fluctuations at the level of single residues. Our approach does not require the acquisition of expensive MD simulations, since it is based on the topology of native contacts derived from a protein's experimental structure; specifically, we made use of normal mode analysis (NMA) [25], which represents, together with principal component analysis (PCA) [26], one of the main protocols employed to identify the most relevant patterns in the large-scale dynamics of proteins. While PCA requires a large set of configurations (for example, from MD trajectories) to build the covariance matrix, NMA can be performed with the sole knowledge of an equilibrium configuration of the system. For this reason, NMA is often used in combination with simplified quadratic models, such as the linearized versions of elastic network models (ENMs) [27]. Another degree of simplification can also be introduced by building coarse-grained (CG) models of the protein, where the atomistic degrees of freedom are replaced by a smaller number of physically relevant representative beads. In spite of this simplicity, the collective, large-scale dynamical features obtained by the NMA of the ENMs of proteins have been shown to be successful to predict experimental B-factors [28] and also conformational changes [29,30].

Given the nature of the ENMs, the proposed pipeline is particularly suited for the study of collective dynamics in globular proteins; ENMs might indeed show limitations for biomolecules whose dynamics is strongly anharmonic, as in the case of intrinsically disordered proteins. For this reason, the validation of the method is herein performed on a set of globular enzymes, namely chymotrypsin-related proteases, for which in-depth analyses of evolutionary relationships and structural similarities are available in the literature [31–34]; in addition, ENM-based NMA has been successfully applied to chymotrypsin-like proteases in previous works, both in the Cartesian space [35,36] and the torsion space [37]. In our approach, normal modes are computed from the β-Gaussian elastic network model of the dataset members [38]. In the β-Gaussian model, each residue is described in a simplified representation as two beads: one corresponds to the C_α atom and represents the main chain, while the second, describing the sidechain, is positioned according to the degrees of freedom of the first bead. An effective quadratic potential energy is used to model the bead fluctuations from the native conformation. We made use of this information to perform a dynamics-based alignment between all pairs of proteins from the dataset; the results from the alignment were used to construct a distance matrix in the space of protein dynamics and to cluster together proteins with similar large-scale motions, thus adding an additional layer of information to clustering procedures based on sequence identity [39,40] or structural similarity [41–43].

Moreover, we developed a way to represent each protein's large-scale normal mode as a vector field on the 3D space. Thanks to this representation, we were able to build a high-dimensional basis set of large-scale protein modes. The basis set is validated by comparison with results from MD simulations, with the perspective of applying this methodology to a dataset comprehensive of a large number of protein classes, differing in structure and function. In this way, common fluctuations between distant proteins can be correlated to the presence of local structural elements, with implications in protein engineering for the design of scaffolds that are able to perform controlled conformational changes in functional enzymes [44,45]. In addition, the large-scale dynamics might serve as a guide to the identification of those patterns where the preservation of a high resolution is of paramount importance in the construction of simplified, multiscale models [46–50] that retain the original dynamics. In particular, by describing at an atomistic level the structural elements identified as important for the desired conformational movements and simultaneously coarse-graining the remainder of the protein, it might be possible to obtain a simplified and computationally inexpensive protein model that shows the conformational dynamics of the high-resolution one.

2. Overview of the Workflow

In our approach, the identification of a common set of conformational motions among different proteins is based on the analysis of their dynamics in a CG representation; from here, a representative set of normal modes is identified through a dynamics-based clustering of the proteins comprising the initial dataset. The selected, representative modes are then orthonormalized and ordered, so as to obtain the final basis set. An overview of the workflow is given in Figure 1 and explained in detail in the following paragraphs.

The starting point is the identification of a set of proteins (Figure 1a). The choice of this dataset is arbitrary and independent of the pipeline; however, the number of proteins that the dataset contains is supposed to be large enough so as to be representative of the families or superfamilies that are included, meaning that the more distant are the members in terms of homology, the larger should be the dataset. This is necessary to ensure the sufficient generality of the resulting basis set of conformational motions.

The selected set of structures is used to run pairwise dynamics-based protein alignments with the ALADYN software developed by some of us [51] (Figure 1b). ALADYN takes two input structures and performs the maximization of a score function that takes into account the spatial superposition of protein regions that have similar motion. The dynamical information is encoded in the low-energy (large-amplitude) eigenvectors obtained from the diagonalization of the interaction matrix M_{ij} of the Hamiltonian function of the β-Gaussian network model:

$$H = \frac{1}{2} \sum_{ij} \delta \vec{x}_i M_{ij} \delta \vec{x}_j \tag{1}$$

where $\delta \vec{x}_i$ is the displacement vector of the i-th bead with respect to the equilibrium configuration. Once the eigenvectors have been obtained, the extent and consistency of the alignment are quantified through the root-mean-squared inner product (RMSIP) between the spaces given by the first 10 modes of each aligned protein. If we call N_i and N_j the total number of residues in the chains of the two aligned proteins, the RMSIP calculation is limited to a subset $q < N_i, N_j$ of marked C_α. These subsets of amino acids are chosen by firstly grouping the amino acids into groups of 10 subsequent ones, then maximizing a single scoring parameter via the standard Metropolis criterion over the space of possible pairs of groups among the two proteins' sequences, as exhaustively explained in [51]. Specifically, the RMSIP is defined as:

$$\text{RMSIP}(\{\vec{v}_l^k\}_i, \{\vec{w}_m^k\}_j) = \text{RMSIP}_{ij} := \sqrt{\frac{1}{10}\sum_{l,m=1}^{10}\left|\sum_{k=1}^{q}\vec{v}_l^k\cdot\vec{w}_m^k\right|^2} \tag{2}$$

The RMSIP $\in [0,1]$ takes on the value of 1 in the case of the perfect correspondence of the spaces and 0 in the case of their complete orthogonality. The quantity (1.0-RMSIP), which still takes values in the interval $[0,1]$, is therefore suitable to define a distance in dynamics between two proteins after alignment. The statistical significance of the alignment, quantified by means of a z-score, is taken into account by weighting the RMSIP by the hyperbolic tangent of the module of the z-score, so as to give more importance to the most reliable results. The distance in dynamics between two aligned proteins i and j is therefore defined as:

$$d_{ij} = 1.0 - \left(\text{RMSIP}_{ij}\cdot\tanh|z_{ij}|\right) \tag{3}$$

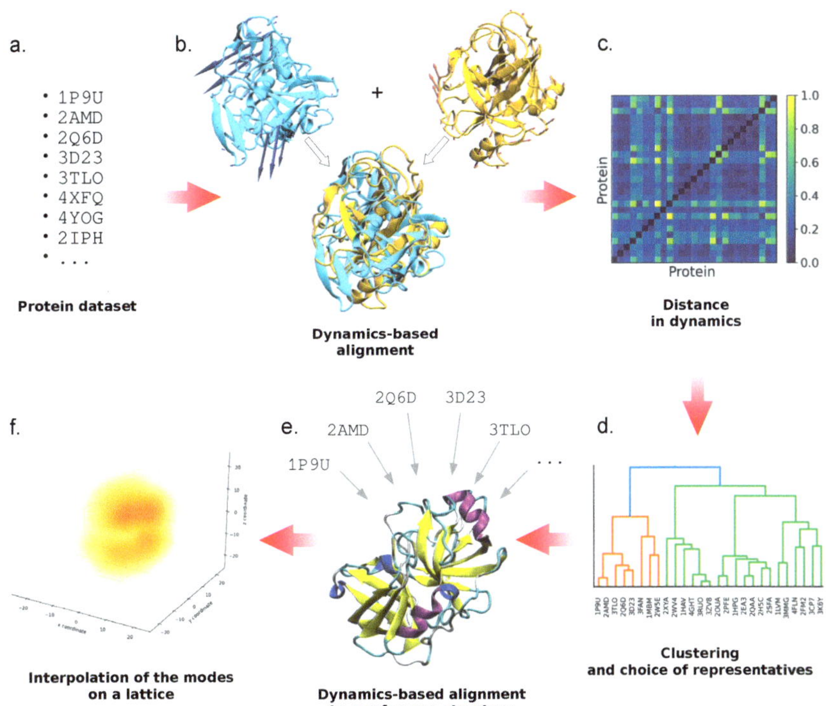

Figure 1. Schematic representation of the workflow proposed. Once the protein dataset is chosen (**a**), dynamics-based alignment is performed between all protein pairs (**b**); the resulting similarity scores (**c**) are used to perform a clustering and to identify one representative protein for each cluster and one for the whole dataset (**d**). All the cluster representatives are dynamically aligned with respect to the latter (**e**), and their normal modes are interpolated on a cubic lattice (**f**). Once orthonormalized and ordered, the latter are used to construct the final basis set.

After all the pairwise alignments between the elements of the dataset are performed, a distance matrix that expresses differences in the large-scale dynamics is obtained (Figure 1c); then, the dataset undergoes hierarchical clustering [52] based on this distance matrix, in order to identify groups of dynamics-related proteins (Figure 1d). The optimal number of clusters is identified from the interplay between *resolution* and *relevance* [53–57]. These two quantities are entropies that are related to each other and depend on the clusterization procedure adopted. We exploited them to select the number of clusters to retain, by considering the smallest number of clusters (hence, the lowest resolution) that gives the highest relevance (Figure 2). Specifically, given a labeling $\hat{s} := (s_1, \ldots, s_\eta)$ (e.g., a clustering) to a sparse dataset made by $N \geq \eta$ data points (in our case, the single proteins in the dataset), the resolution is defined as an entropy $\hat{H}_{res}$ representing the relative amount of information loss in the process:

$$\hat{H}_{res}[\hat{s}] := -\sum_s p_s \cdot \log_2(p_s) \quad p_s := \frac{k_s}{N} \tag{4}$$

where k_s is the number of data points that fall into the same cluster s. It has been proven [54] that $\hat{H}_{res}$ increases monotonically with the number of clusters, in accordance with the idea that the coarser is our clustering, the more information we lose. On the other hand, the relevance $\hat{H}_{rel}$ is defined as:

$$\hat{H}_{rel}[\hat{k}] := -\sum_k \frac{k \cdot m_k}{N} \cdot \log_2\left(\frac{k \cdot m_k}{N}\right) \tag{5}$$

where m_k is the number of clusters containing the same amount $k = 0, \ldots, N$ of data points, for a given clustering process. By choosing the lowest resolution value corresponding to the largest relevance (Figure 2), we can rely on the most compact clusterization (thus increasing the statistics within each cluster) that preserves the highest empirical information content.

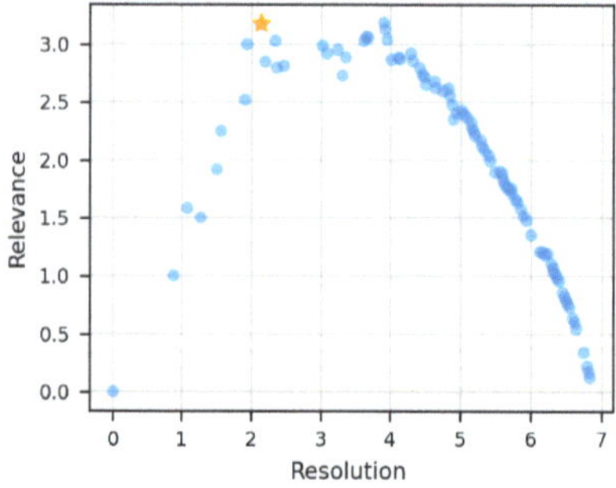

Figure 2. Resolution–relevance curve used to determine the optimal number of clusters in the dynamics-based clusterization of the protein dataset. Each point corresponds to a different number of clusters. The optimal subdivision, indicated with an orange star, corresponds to 9 clusters.

Once the optimal number of clusters is derived, protein representatives of each cluster are identified as the cluster centroids, namely the proteins with the shortest distance to every other protein of the cluster itself. In addition, a representative for the whole dataset is selected as the protein with the most characteristic dynamics, expressed in terms of the lowest distance with respect to all the other dataset members. The other protein structures are then dynamically aligned to this one with ALADYN, so as to have a consistent orientation in space (Figure 1e).

From an ENM representation of each of these newly oriented structures, normal modes are computed. In order to facilitate the comparison between modes belonging to proteins with a different sequence length, the first five reoriented normal modes of the cluster representatives are placed on a cubic lattice and interpolated on the grid points so as to obtain a smooth vector field (Figure 1f). In this way, we move from comparing the $3N$-dimensional modes of different proteins (where N is the number of residues, different for each protein), to comparing vector fields defined on identical 3D lattices having the same dimension. More details on the lattice construction and interpolation are given in Section 3. Proteins belonging to the dataset employed in this work, despite displaying a range of sequence length and radius of gyration, do not grandly differ in size; therefore, the modes interpolated on the lattice can be directly compared. However, it might be the case that the dataset includes proteins with very different size; this would require a rescaling of the protein coordinates before the interpolation on the lattice, so as to compare motions occupying similar volumes in space.

The interpolated modes are orthonormalized using the Gram–Schmidt algorithm [58]. The components of the basis are finally ordered according to decreasing entropy, considered as a measure of their degree of collectivity. The entropy S of a mode k is defined as:

$$S_k = -\frac{\sum_i \phi_i^k \ln \phi_i^k}{\ln N},\tag{6}$$

where N is the number of lattice sites and ϕ_i^k is the square modulus of the k-th mode on the lattice site i. S_k takes a maximum value of 1 if the mode is delocalized on all the lattice sites and a minimum value of 0 if the mode is localized on a single site.

The final set of orthonormalized and ordered vector spaces represents the basis of protein dynamics. In the next section, technical details of the methods employed are presented.

3. Materials and Methods

3.1. Preprocessing of the Dataset

A dataset of 116 chymotrypsin-related proteases, for which structural experimental information is available, was selected. This dataset is based on the one used in [34], from which proteins with sequence identity $> 70\%$ were removed. The dataset comprises serine proteases from bacteria, eukaryotes, archaea, and viruses, in addition to chymotrypsin-related cysteine proteases from positive-strand RNA viruses. The full list of the proteins' PDB IDs is given in Table S1. The structures were downloaded from the Protein Data Bank, and the coordinate files were cleaned-up from heteroatoms, from copies of the protein in the crystallographic cell, and from residue-configurations with low occupancy. The position of missing atoms was rebuilt and the protein conformations were optimized using the software FoldX 4 [59]. Non-terminal missing residues were modeled with MODELLER [60,61]. An analysis of the first 3 normal modes for each protein was run using an elastic network model with a cutoff of 10 Å, in order to identify the problematic cases in which the flexible protein termini impaired the analysis of the motion of the protein core. Such analysis was conducted by visual inspection of the modes on the protein structures. In those cases, flexible tails were not considered in the following analyses, which thus focused on globular structures. Moreover, in the case of multi-domain structures, only the domain known to have protease activity was retained.

3.2. Dynamics-Based Alignment and Clustering

The dynamics-based alignment of all the pairs of protein structures was performed with the ALADYN software [51], developed by some of us, using as input the cleaned coordinates files. From the resulting alignment scores, clustering of the structures was performed with the Python library SciPy, using the ward linkage method. The calculation of relevance and resolution, used to identify the optimal number of clusters, was performed with an in-house script.

3.3. Lattice Interpolation and Basis Construction

Normal modes of each protein of the dataset were computed with an in-house code. The first 5 reoriented normal modes of the cluster representatives were placed on a cubic lattice, with a lattice constant of 1 Å (for a total of 45 modes, namely vector fields). The vector on each protein C_α was translated on the nearest lattice grid point. The mode vectors were interpolated on the lattice in order to create a smooth vector field (Figure 1), using Gaussian functions with $\sigma = 0.8$ Å and truncated at a distance of 2 Å. This distance is slightly smaller than the lowest spatial distance between two C_α atoms to make sure that the vector coming from the original protein mode is not spuriously modified during interpolation. The chosen value of σ ensures that, in correspondence with the cutoff, the mode field is close to zero. The resulting vector at each grid point *ijk* is the sum of the mode fields centered on the nearby C_α grid points, calculated at *ijk*, within the cutoff. Eventually, the orthonormalization and ordering of the modes were performed with Python scripts.

3.4. Molecular Dynamics Simulations

Molecular dynamics simulations have been performed on the representatives of each cluster, using the software Gromacs 2019 [62]. The proteins were described with the Amber99sb-ildn force field [63], and the TIP3P model [64] was used for water molecules. Sodium and chloride ions were added at a concentration of 0.15 M and balanced so as to neutralize the charge in the simulation box. All systems were energy minimized for 100 steps by steepest descent. The solvent was then equilibrated for 500 ps with positional restraints on the protein heavy atoms, using a force constant of 1000 kJ·mol^{-1}·nm^{-2}. MD simulations were carried out in the NPT ensemble for 250 ns for each system. Protein and solvent were coupled separately to a 300 K heat bath with a coupling constant of 0.1 ps, using the velocity-rescaling thermostat [65]. The systems were isotropically pressure-coupled at 1 bar with a coupling constant of 2.0 ps, using the Parrinello–Rahman barostat [66]. The application of the LINCS [67] algorithm on hydrogen-containing bonds allowed for an integration time step of 2 fs. Short-range electrostatic and Lennard–Jones interactions were calculated within a cut-off of 1.0 nm, and the neighbor list was updated every 10 steps. The particle mesh Ewald (PME) method was used for the long-range electrostatic interactions [68], with a grid spacing of 0.12 nm.

The calculation of the root-mean-squared fluctuations from the trajectory coordinates was performed on the protein C_α atoms using the Gromacs tool *gmx rmsf*. The dynamic cross-correlation was computed with a Python script, using the library MDTraj [69]. Plots were produced with Python libraries, and protein images were rendered with VMD [70].

4. Results and Discussion

4.1. Overview of the Protein Dataset

Proteases are enzymes catalyzing the reaction of the hydrolysis of peptide bonds. The independent evolutionary origin of these enzymes [71] is reflected in their large variety of sizes, shapes, and specificity [72]. In this work, we focus on a specific superfamily, namely the chymotrypsin-related proteases. The latter share a common structure with two β-barrel-like domains accommodating the binding site (Figure 3); however, the size and structural completeness of the β-barrels and the length of the turns and loops connecting the sheets greatly vary. The result of this structural variability is a range of sequence lengths and protein sizes among the 116 proteins included in our dataset (Figure S1). The proteolytic reaction is performed by a catalytic triad of residues, located between the β-barrels. The type of amino acid playing the role of nucleophile in the mechanism of catalysis determines the class of proteases: in the serine proteases, the catalytic triad contains His, Asp/Glu, and Ser residues [73]; in the cysteine proteases, the triad is composed of His, Asp/Glu, and Cys or of a dyad of His and Cys residues [74].

The classification used in the remainder of the paper is based on MEROPS, a hierarchical classification scheme for proteases [75,76]. In the MEROPS database, chymotrypsin-related proteases constitute the PA clan, which contains 9 families of cysteine proteases (representing proteases of positive-strand RNA viruses) and 14 families of serine proteases (representing proteolytic enzymes from eukaryotes, bacteria, some DNA viruses, and eukaryotic positive-strand RNA viruses). Families are defined on the basis of sequence similarity and/or resemblance of the folds among their protein members. However, experimental structural information is available for a limited number of these families; therefore, not all of them are represented in the dataset employed in this work.

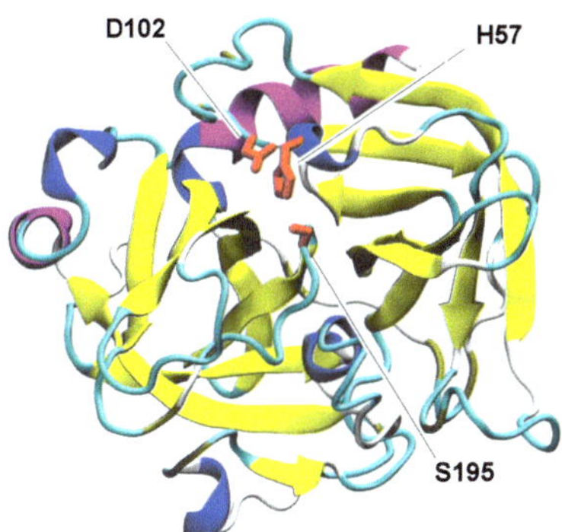

Figure 3. Cartoon representation of chymotrypsin from *Bos taurus* (PDB ID: 2CGA). Colors are used to differentiate the structural elements; in particular, the two β-barrels are distinguishable in yellow. The catalytic triad is represented in licorice and colored in red.

4.2. Results of the Dynamics-Based Alignment

We performed an alignment based on the dynamical information entailed into the first 10 lowest frequency modes obtained by the NMA on the β-Gaussian network model of each pair of proteins in the dataset. The alignment consists of the optimization of a score function that maximizes the RMSIP of the two sets of normal modes. For each pair of dynamically aligned proteins, matching regions in the two structures are identified as the subset of residues giving the best overlap. The number of residues belonging to these cores shows great variability (Figure S2), and their RMSD values range from 0.6 to 4.0 Å; these results are indicative of heterogeneity in dynamics within the dataset.

The distance matrix obtained from the pairwise dynamics-based alignments of all proteins of this dataset is used as a measure of similarity in dynamics. This can be compared to the MEROPS classification by computing the average distance between protein pairs that fall into the same family. Following such a procedure, it is apparent that the average distance in dynamics is lower within each family, with respect to the total average (Figure 4). In other words, proteins belonging to the same family are significantly closer in dynamics than they are to members of other families.

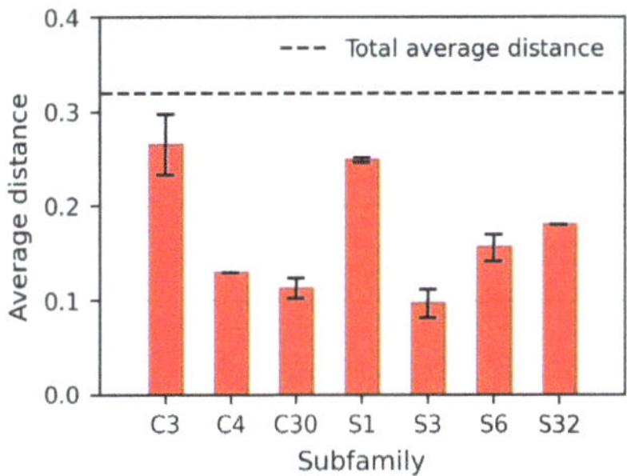

Figure 4. Average distances (in terms of dynamics) between proteins of the dataset belonging to the same family. Only those subfamilies including more than one representative member are displayed here. The histograms show that proteins are significantly closer in dynamics within the same family than they are to members of other families.

The distance matrix is used as the input for the division of the dataset into dynamically homogeneous protein clusters. The outcome of the hierarchical clustering is graphically expressed by the dendrogram in Figure S3. On the basis of the resolution–relevance plot, nine clusters were identified (Figure 2); this corresponds to a threshold of ≈0.58 in the clustering dendrogram. The resulting clusters appear to be quite homogeneous in terms of protease classification (Figure S4). Importantly, the dynamics-based clustering automatically tends to group proteins belonging to the same subfamily. Figure 5a shows that in most of the cases (17 of the 19 subfamilies represented in the dataset), all the members of each subfamily fall into the same cluster, thus suggesting that these proteins share a similar conformational dynamics and strengthening the idea of homogeneity in dynamics between homologous proteins [77,78]. On the other hand, each cluster groups several subfamilies, and only 4 clusters out of 9 include proteins belonging to only one subfamily (Figure 5b). Therefore, the clustering procedure proves able to effectively group different protein subfamilies that, despite the different evolutionary origin, share similar dynamics.

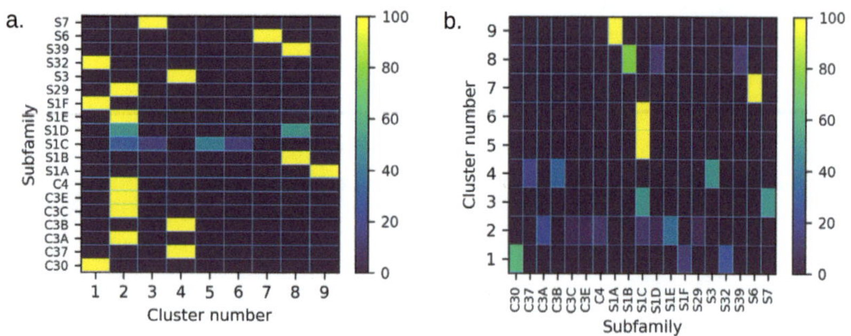

Figure 5. (**a**) Distribution of the members of each subfamily among the different clusters, expressed as a percentage with respect to the total number of members of the subfamily. In (**b**), each row represents the content of each cluster classified on the basis of the function (in percentage, with respect to the total population of the cluster). The results show that the dynamics-based clustering automatically tends to group proteins belonging to the same subfamily.

4.3. Comparison between the Dynamics-Based and the Structure-Based Clustering

We compared the results from the dynamics-based clustering on the proteases of the PA clan with the structure-based distance tree calculated in the work of Mönttinen et al. [34]. There, the authors identified a common structural core of 72 residues for the set of PA clan proteases taken into account; according to the structural similarities of this common core, they built a distance tree between the members of the dataset. Five different clusters were identified, contrary to the nine cluster found in this work.

Despite the two different approaches, the results present several similarities, showing a close relation between structure and dynamics. The S1A subfamily, which includes both bacterial and eukaryotic proteases, forms a clearly distinct and compact cluster both in terms of structure and dynamics. On the other hand, the S1D subfamily, which includes bacterial proteases, is split into two different groups in terms of structure, as well as dynamics: in both cases, the S1D *Achromobacter* protease I (1ARB) is close to the bacterial S1B proteases, while the S1D protease AL20 of *Nesterenkonia abyssinica* (3CP7) is close to the members of the bacterial S1E subfamily. This difference between members of the S1D subfamily has been explained on the basis of the different evolutionary history of the bacteria in which they are expressed [34].

Another common feature emerging from the two clustering approaches is the similarity between the S39 subfamily of positive-strand RNA viruses and the bacterial S1B proteases; interestingly, such a degree of similarity is higher than between S39 and the other viral proteases, as already reported on the basis of structural comparisons [79]. Moreover, the bacterial S6 family forms an independent group in both clustering approaches. This peculiarity has been attributed to the presence of a long β-stalk structure at the C-terminus (Figure S5), which is absent in all the other proteases of the PA clan [34,80]; the protease domain alone, instead, shares high structural similarity with that of the S1A subfamily. However, the β-stalk domain was cut before the dynamics-based alignment, meaning that our analysis of the dynamics of the S6 protease domain alone is able to distinguish this subfamily from the other members of the PA clan.

Importantly, the two types of clustering present also some differences. In the case of the structure-based analysis, the cysteine proteases tend to be grouped together; however, in the dynamics-based alignment, the similarity is only at the level of one of the two large groups into which the dataset is divided, as evident from the dendrogram in Figure S4. Within this group, C families are mixed with S families and appear to be more distributed among different clusters than in the distance tree built on the basis of the structural features. This is indicative of a clear differentiation of the C proteases in terms of dynamics, despite their structural similarity in the protein core. This can be explained not only by the fact that different classes of C proteases are involved in the processing of different viral polyproteins (therefore, requiring adaptation to the substrate), but also because some of them have additional functions, playing the role of inhibitors of host cell protein synthesis [81]. Another difference regards the heat-shock proteases S1C, which include proteins from bacteria, chloroplasts, and mitochondria; even though structurally similar in the proteolytic core, members of this subfamily appear very scattered in the dynamics-based clustering. Specifically, the observed similarities in the dynamics accentuates the structural relatedness already observed between some eukaryotic S1C proteases and different viral protease subfamilies, in that these similarities are stronger than the similarity within the S1C subfamily itself. This relatedness has been previously explained on the basis of exchanges of protease genes between eukaryotic viruses and their hosts [34].

In the structure-based distance tree, proteases from flavivirus (families S29 and S7) and from togavirus (family S3) are grouped together, even though the two viruses belong to different families; on the other hand, S29/S7 and S3 are placed in different clusters when their dynamics is included in the analysis. This distinction might arise from the difference in function: the S3 protein togavirin, in fact, does not only function as a viral protease, but plays also the structural role of the capsid protein of the virus [82]. S29 and S7 proteases, on the other hand, possess only proteolytic function and do not work as structural components.

Overall, the inclusion of dynamics in the comparison of the proteases from the PA clan adds therefore an additional level of classification, which seems appropriate to bridge structural and functional similarities.

4.4. Creation and Validation of the Basis Set of the High-Dimensional Space of Protein Dynamics

The representative proteins of the nine clusters are identified by the PDB codes: 3D23, 1HPG, 2YOL, 1VCP, 3QO6, 1L1J, 1WXR, 4JCN, and 4I8H. Their structures are represented in Figure S6. Protein 1GDQ was chosen as the reference structure of the whole dataset, against which the other representatives are dynamically aligned prior to lattice interpolation of their normal modes (see Section 3). In the latter, the oriented protein modes are placed and interpolated on a cubic lattice, orthonormalized, and finally, ordered. The interpolation on the grid allows us to easily compare the dynamics of any pair of proteins, irrespective of the number of residues. For instance, modes from proteins with a different number of C_α cannot be directly compared in terms of scalar products, while different vector fields on the grid have the same dimensionality.

We investigated the quality of the orthonormalized modes as a basis set for the dynamics of the whole dataset, by computing the overlap between the spaces given by the protein modes and by the basis. To this aim, the RMSIP was computed between the space spanned by the first five modes of each protein in the dataset (after their interpolation on the lattice) and the first n components of the basis. For each protein, the components of the basis are ordered so as to maximize the RMSIP with the protein modes. The resulting

RMSIP for each protein is plotted in Figure 6a as a function of the number n of basis vectors considered for the calculation of the RMSIP. From the distribution of the values attained when using the full basis set (45 vector fields), the RMSIP is greater than 0.5 for $\approx$94% of the proteins, showing in those cases a good agreement between the dynamics of the protein and the one expressed by the basis [83]. The agreement is excellent (RMSIP > 0.7) for $\approx$61% of the proteins; therefore, we can conclude that the identified basis is indeed able to describe with good generality the large-scale conformational dynamics of the dataset. For each protein, we also computed the normalized RMSIP, by dividing each value of the RMSIP with the value obtained with the use of the full basis set. The normalized RMSIP curves show that, for each dataset member, as few as 15 basis components are sufficient to reproduce 80% of the dynamics that would be attained with the use of the full basis set (Figure 6b); however, such components differ from protein to protein, meaning that there are no vector fields in the basis that can be considered more essential than others. This suggests that a further reduction in the dimension of the basis set would lead to a loss of generality in the description of the dynamics of this class of proteins.

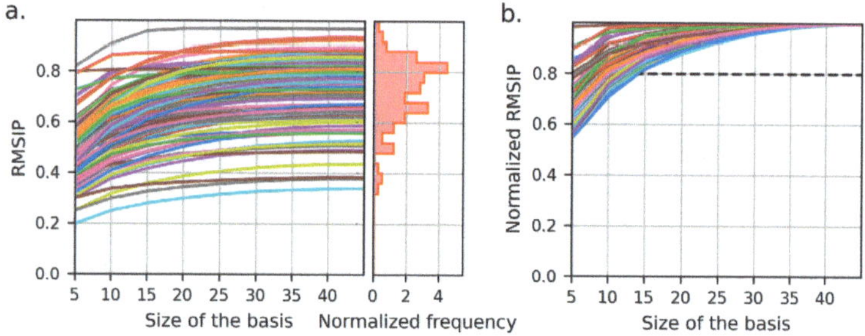

Figure 6. (**a**) Root-mean-squared inner product (RMSIP) between the subspaces spanned by the first 5 modes of each protein and the first n basis vectors, as a function of the basis size n. Each line corresponds to one protein of the dataset. The histogram on the right represents the distribution of the RMSIP values attained when the full basis is used. The RMSIP shows a good overlap of the subspaces (RMSIP > 0.5) for $\approx$94% of the proteins. (**b**) RMSIP normalized with respect to the value attained from the use of the full basis. For each dataset member, as few as 15 basis components are sufficient to reproduce 80% of the dynamics that would be attained with the use of the full basis set.

4.5. Comparison with MD Simulations

In order to better assess the ability of the basis to reproduce the general dynamics of chymotrypsin-like proteases, we performed MD simulations of four proteins belonging to the same family and compared the per-residue fluctuations emerging from the simulations with those obtained by filtering the trajectory along the vectors of the basis; a good agreement would be indicative of the ability of the basis to describe the large-scale dynamics of the protein. Two of the proteins used as a test-case belong to the dataset; these are 1EKB [84] and 1NPM [85], eukaryotic proteases belonging to the S1A subfamily. The other two proteins, 4YOG [86] and 3W94 [87], are external to the dataset and, as such, have not been used to define the basis. 4YOG is a C30 protease from the bat coronavirus HKU4, while 3W94 is an S1A enteropeptidase. These two proteins have been included here in order to test the generality of the identified basis for the description of the dynamics of the PA clan, independently of the specific members of the initial dataset.

For each of the four proteins we compared the root-mean-squared fluctuations (RMSF) as computed from the simulation and as computed from the same trajectory filtered along the "modes" given by the backmapping of the protein structure on the basis vectors. The comparison shows a good qualitative agreement (Figures 7 and S7), in particular in correspondence with all the secondary structure elements. In the unstructured regions, the comparison is slightly less accurate; this is particularly true for long loops, which are more sensitive to the limitations of the ENM and of the NMA employed to define the modes of the basis, since both assume small-amplitude fluctuations from a well-defined reference structure. From the two sets of trajectories, namely the original MD simulations and the filtered ones, we also computed the dynamic cross-correlation matrices (Figures S8 and S9), which give a measure of the degree of correlation between each pair of C_α atoms in terms of fluctuations from their average position. When comparing the original and filtered trajectories, the intensity of the resulting correlations are different, with higher correlations/anti-correlations emerging from the trajectory filtered on the basis; however, the patterns of correlation are strikingly similar between the two trajectories for all four proteins. In addition, we computed the RMSIP between the first n modes obtained from the PCA of the MD simulation and of the filtered trajectory, where n is the number of components that capture 80% of the variance in the original simulation (Table S2); in all cases, the results show a good overlap of the two subspaces, with RMSIP > 0.5. Therefore, the basis set appears to be able to describe the relevant large-scale dynamics of the considered protein systems.

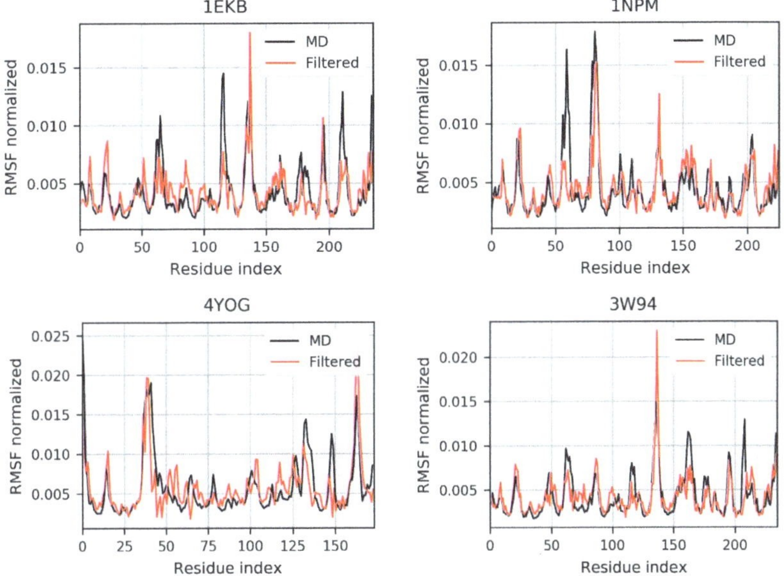

Figure 7. Root-mean-squared fluctuations (RMSF) of the C_α atoms, normalized with respect to their sum, computed on proteins belonging to the initial dataset (1EKB, 1NPM) and external to it (4YOG, 3W94). The shaded areas correspond to structured regions, identified with the DSSP algorithm [88,89]. The comparison shows a good qualitative agreement, particularly in correspondence with secondary structure elements.

5. Conclusions

In this work, we proposed a workflow for the identification of common large-scale conformational motions in a set of proteins. Specifically, we performed a dynamics-based clusterization of 116 chymotrypsin-related proteases, belonging to the PA clan, and compared the resulting clusters to the MEROPS classification and to a more recent structure-based classification of the same dataset of proteases. The clustering based on the dynamics adds interesting information to that known on the basis of structural and evolutionary relationships between the members of the protein family, thus facilitating the interpretation of dynamics as a bridge between protein structure and function. In addition, we used NMA and the β-GNM to build a basis set of vectors of the high-dimensional space of the PA clan large-scale dynamics and tested the basis set to demonstrate that it is sufficiently complete to describe the main large-scale dynamical features of the members of the dataset. The basis set of conformational motions was also successfully validated by comparison with results from MD simulations of proteins internal and external to the initial dataset.

In this regard, the method proved to deal particularly well with the conformational dynamics of structured regions; loops and disordered regions are by definition challenging to describe with an ENM, which is able to reproduce only small-amplitude fluctuations with respect to a well-defined reference structure; the dynamics of such regions, however, is qualitatively different from the functional one of the structured part, which is the one responsible for carrying out the biological function in the proteins under examination. Additionally, we note that the dataset employed contained only a number of proteins belonging to the family of chymotrypsin-related proteases: a larger dataset is expected to lead to more general results; however, the number of proteins included was limited by the availability of experimental structures and by the choice to remove proteins with too high sequence identity. The natural development of the methodology presented and discussed in this work is its application to a larger dataset of proteins, comprehensive of multiple enzyme superfamilies, with the aim of building a basis set of conformational motions that represents a general vocabulary of proteins' common dynamics. Once mapped on a protein structure, the basis components can help to identify the most common—but diverse among each other—movements that better describe the common large-scale dynamics of the proteins belonging to the dataset. The dynamics of any protein not belonging to the initial set can be projected on the basis, so as to describe it in terms of a few general movements, thus facilitating the comparison between the dynamical features of different proteins. In addition, the method can be employed to identify those common structural signatures that characterize the dynamics encoded in the basis components and relate them to specific biological functions.

Supplementary Materials: The following supporting information can be downloaded at: https://www.mdpi.com/article/10.3390/app12147157/s1, Figure S1: Histograms of the sequence length (a) and radius of gyration (b) of the proteins in the dataset. Figure S2: Histograms of the number of residues belonging to the superimposed protein cores, defined from the dynamics-based alignment of each pair of proteins from the dataset. Figure S3: Dendrogram resulting from the hierarchical clustering, performed on the basis of the distance in dynamics between the dataset elements. The labels represents the PDB IDs, and colors are used to differentiate the clusters. Figure S4: Dendrogram resulting from the hierarchical clustering, performed on the basis of the distance in dynamics between the dataset elements. The labels represents the protease subfamily of each protein, and colors are used to differentiate the clusters. Figure S5: (a) Full structure of the 1WXR protease from subfamily S6, displaying the long β-stalk domain at the C-terminus. (b) Structural alignment of 1WXR (in cyan) and 4I8H from subfamily S1A (in orange), showing the similarity of their protein core. Figure S6: Structure of the representatives of each protein cluster, resulting from the dynamics-based alignment. The color corresponds to the type of secondary structure element: β-sheets in yellow, α-helices in magenta, 3–10 helices in blue and loops in cyan. Figure S7: Scatter plots of the root-mean-square fluctuation (RMSF) values, computed on the C_α atoms, from the MD simulations of the protein and from the same trajectories filtered on the basis set. ρ indicates the value of Pearson Coefficient computed between the two sets of fluctuations. All cases show satisfactory results. Figure S8: Cross-correlation computed from the simulations of the proteins 1EKB and 1NPM, both on the original and filtered trajectories. Both proteins belong to the dataset. Figure S9: Cross-correlation computed from the simulations of the two proteins 4YOG and 3W94, both on the original and filtered trajectories. The two proteins are not part of the dataset from which the basis set is derived. Table S1: List of the PDB IDs of the proteins comprising the dataset. Table S2: RMSIP computed between the first n modes obtained from the PCA of the MD simulation and of the filtered trajectory, where n is the number of components that capture the 80% of the variance in the original trajectory. The results show a good overlap of the two subspaces in all the simulated systems.

Author Contributions: Conceptualization: R.P.; methodology, data collection, and analysis: T.T. and M.R.; writing—original draft preparation: T.T. and G.M.; writing—review and editing: T.T., G.M., M.R. and R.P.; supervision: R.P.; funding acquisition: R.P. All authors have read and agreed to the published version of the manuscript.

Funding: This project received funding from the European Research Council (ERC) under the European Union's Horizon 2020 research and innovation program (Grant 758588).

Institutional Review Board Statement: Not applicable.

Informed Consent Statement: Not applicable.

Data Availability Statement: The raw data produced and analyzed in this work are freely available on the Zenodo repository at https://doi.org/10.5281/zenodo.6669245.

Acknowledgments: The authors thank Roberto Menichetti for a critical and insightful reading of the manuscript.

Conflicts of Interest: The authors declare no conflict of interest.

References

1. Berendsen, H.J.; Hayward, S. Collective protein dynamics in relation to function. *Curr. Opin. Struct. Biol.* **2000**, *10*, 165–169. [CrossRef]
2. Henzler-Wildman, K.A.; Lei, M.; Thai, V.; Kerns, S.J.; Karplus, M.; Kern, D. A hierarchy of timescales in protein dynamics is linked to enzyme catalysis. *Nature* **2007**, *450*, 913–916. [CrossRef] [PubMed]
3. Narayanan, C.; Bafna, K.; Roux, L.D.; Agarwal, P.K.; Doucet, N. Applications of NMR and computational methodologies to study protein dynamics. *Arch. Biochem. Biophys.* **2017**, *628*, 71–80. [CrossRef]
4. Ma, B.; Kumar, S.; Tsai, C.J.; Nussinov, R. Folding funnels and binding mechanisms. *Protein Eng.* **1999**, *12*, 713–720. [CrossRef] [PubMed]
5. Nussinov, R.; Ma, B. Protein dynamics and conformational selection in bidirectional signal transduction. *BMC Biol.* **2012**, *10*, 2. [CrossRef]
6. Koshland, D., Jr. Application of a theory of enzyme specificity to protein synthesis. *Proc. Natl. Acad. Sci. USA* **1958**, *44*, 98. [CrossRef]
7. Paul, F.; Weikl, T.R. How to distinguish conformational selection and induced fit based on chemical relaxation rates. *PLoS Comput. Biol.* **2016**, *12*, e1005067. [CrossRef]
8. Yang, L.Q.; Sang, P.; Tao, Y.; Fu, Y.X.; Zhang, K.Q.; Xie, Y.H.; Liu, S.Q. Protein dynamics and motions in relation to their functions: Several case studies and the underlying mechanisms. *J. Biomol. Struct. Dyn.* **2014**, *32*, 372–393. [CrossRef]
9. Hensen, U.; Meyer, T.; Haas, J.; Rex, R.; Vriend, G.; Grubmüller, H. Exploring protein dynamics space: The dynasome as the missing link between protein structure and function. *PLoS ONE* **2012**, *7*, e33931. [CrossRef]
10. Kern, D.; Zuiderweg, E.R. The role of dynamics in allosteric regulation. *Curr. Opin. Struct. Biol.* **2003**, *13*, 748–757. [CrossRef]
11. Zhang, Y.; Doruker, P.; Kaynak, B.; Zhang, S.; Krieger, J.; Li, H.; Bahar, I. Intrinsic dynamics is evolutionarily optimized to enable allosteric behavior. *Curr. Opin. Struct. Biol.* **2020**, *62*, 14–21. [CrossRef] [PubMed]
12. Liang, Z.; Verkhivker, G.M.; Hu, G. Integration of network models and evolutionary analysis into high-throughput modeling of protein dynamics and allosteric regulation: Theory, tools and applications. *Briefings Bioinform.* **2020**, *21*, 815–835. [CrossRef] [PubMed]
13. Balsera, M.A.; Wriggers, W.; Oono, Y.; Schulten, K. Principal component analysis and long time protein dynamics. *J. Phys. Chem.* **1996**, *100*, 2567–2572. [CrossRef]
14. Stein, S.A.M.; Loccisano, A.E.; Firestine, S.M.; Evanseck, J.D. Principal components analysis: A review of its application on molecular dynamics data. *Annu. Rep. Comput. Chem.* **2006**, *2*, 233–261.
15. Kmiecik, S.; Kouza, M.; Badaczewska-Dawid, A.E.; Kloczkowski, A.; Kolinski, A. Modeling of protein structural flexibility and large-scale dynamics: Coarse-grained simulations and elastic network models. *Int. J. Mol. Sci.* **2018**, *19*, 3496. [CrossRef]
16. Marsh, J.A.; Teichmann, S.A. Parallel dynamics and evolution: Protein conformational fluctuations and assembly reflect evolutionary changes in sequence and structure. *BioEssays* **2014**, *36*, 209–218. [CrossRef]
17. Zou, T.; Risso, V.A.; Gavira, J.A.; Sanchez-Ruiz, J.M.; Ozkan, S.B. Evolution of conformational dynamics determines the conversion of a promiscuous generalist into a specialist enzyme. *Mol. Biol. Evol.* **2015**, *32*, 132–143. [CrossRef]
18. Narayanan, C.; Bernard, D.N.; Bafna, K.; Gagné, D.; Chennubhotla, C.S.; Doucet, N.; Agarwal, P.K. Conservation of dynamics associated with biological function in an enzyme superfamily. *Structure* **2018**, *26*, 426–436. [CrossRef]
19. Zhang, S.; Li, H.; Krieger, J.M.; Bahar, I. Shared signature dynamics tempered by local fluctuations enables fold adaptability and specificity. *Mol. Biol. Evol.* **2019**, *36*, 2053–2068. [CrossRef]
20. Mikulska-Ruminska, K.; Shrivastava, I.; Krieger, J.; Zhang, S.; Li, H.; Bayır, H.; Wenzel, S.E.; VanDemark, A.P.; Kagan, V.E.; Bahar, I. Characterization of differential dynamics, specificity, and allostery of lipoxygenase family members. *J. Chem. Inf. Model.* **2019**, *59*, 2496–2508. [CrossRef]

21. Gaur, N.K.; Ghosh, B.; Goyal, V.D.; Kulkarni, K.; Makde, R.D. Evolutionary conservation of protein dynamics: Insights from all-atom molecular dynamics simulations of 'peptidase'domain of Spt16. *J. Biomol. Struct. Dyn.* **2021**, 1–13. [CrossRef] [PubMed]
22. Maguid, S.; Fernandez-Alberti, S.; Echave, J. Evolutionary conservation of protein vibrational dynamics. *Gene* **2008**, *422*, 7–13. [CrossRef] [PubMed]
23. Velázquez-Muriel, J.A.; Rueda, M.; Cuesta, I.; Pascual-Montano, A.; Orozco, M.; Carazo, J.M. Comparison of molecular dynamics and superfamily spaces of protein domain deformation. *BMC Struct. Biol.* **2009**, *9*, 6. [CrossRef]
24. Pearl, F.; Todd, A.; Sillitoe, I.; Dibley, M.; Redfern, O.; Lewis, T.; Bennett, C.; Marsden, R.; Grant, A.; Lee, D.; et al. The CATH Domain Structure Database and related resources Gene3D and DHS provide comprehensive domain family information for genome analysis. *Nucleic Acids Res.* **2005**, *33*, D247–D251. [CrossRef] [PubMed]
25. Levitt, M.; Sander, C.; Stern, P.S. Protein normal-mode dynamics: Trypsin inhibitor, crambin, ribonuclease and lysozyme. *J. Mol. Biol.* **1985**, *181*, 423–447. [CrossRef]
26. David, C.C.; Jacobs, D.J. Principal Component Analysis: A Method for Determining the Essential Dynamics of Proteins. In *Protein Dynamics: Methods and Protocols*; Humana Press: Totowa, NJ, USA, 2014; pp. 193–226.
27. Tirion, M.M. Large Amplitude Elastic Motions in Proteins from a Single-Parameter, Atomic Analysis. *Phys. Rev. Lett.* **1996**, *77*, 1905–1908. [CrossRef]
28. Zheng, W. Anharmonic normal mode analysis of elastic network model improves the modeling of atomic fluctuations in protein crystal structures. *Biophys. J.* **2010**, *98*, 3025–3034. [CrossRef]
29. Dobbins, S.E.; Lesk, V.I.; Sternberg, M.J.E. Insights into protein flexibility: The relationship between normal modes and conformational change upon protein-protein docking. *Proc. Natl. Acad. Sci. USA* **2008**, *105*, 10390–10395. [CrossRef]
30. Delarue, M.; Sanejouand, Y.H. Simplified Normal Mode Analysis of Conformational Transitions in DNA-dependent Polymerases: The Elastic Network Model. *J. Mol. Biol.* **2002**, *320*, 1011–1024. [CrossRef]
31. Gorbalenya, A.E.; Donchenko, A.P.; Blinov, V.M.; Koonin, E.V. Cysteine proteases of positive strand RNA viruses and chymotrypsin-like serine proteases: A distinct protein superfamily with a common structural fold. *FEBS Lett.* **1989**, *243*, 103–114. [CrossRef]
32. Di Cera, E. Serine proteases. *IUBMB Life* **2009**, *61*, 510–515. [CrossRef] [PubMed]
33. Laskar, A.; Rodger, E.J.; Chatterjee, A.; Mandal, C. Modeling and structural analysis of PA clan serine proteases. *BMC Res. Notes* **2012**, *5*, 1–11. [CrossRef] [PubMed]
34. Mönttinen, H.A.; Ravantti, J.J.; Poranen, M.M. Structural comparison strengthens the higher-order classification of proteases related to chymotrypsin. *PLoS ONE* **2019**, *14*, e0216659. [CrossRef] [PubMed]
35. Ma, W.; Tang, C.; Lai, L. Specificity of trypsin and chymotrypsin: Loop-motion-controlled dynamic correlation as a determinant. *Biophys. J.* **2005**, *89*, 1183–1193. [CrossRef]
36. Sola, R.J.; Griebenow, K. Influence of modulated structural dynamics on the kinetics of α-chymotrypsin catalysis: Insights through chemical glycosylation, molecular dynamics and domain motion analysis. *FEBS J.* **2006**, *273*, 5303–5319. [CrossRef]
37. Dauber-Osguthorpe, P.; Osguthorpe, D.J.; Stern, P.S.; Moult, J. Low frequency motion in proteins: Comparison of normal mode and molecular dynamics of streptomyces griseus protease A. *J. Comput. Phys.* **1999**, *151*, 169–189. [CrossRef]
38. Micheletti, C.; Carloni, P.; Maritan, A. Accurate and efficient description of protein vibrational dynamics: Comparing molecular dynamics and Gaussian models. *Proteins Struct. Funct. Bioinform.* **2004**, *55*, 635–645. [CrossRef]
39. Li, W.; Fu, L.; Niu, B.; Wu, S.; Wooley, J. Ultrafast clustering algorithms for metagenomic sequence analysis. *Briefings Bioinform.* **2012**, *13*, 656–668. [CrossRef]
40. Gabler, F.; Nam, S.Z.; Till, S.; Mirdita, M.; Steinegger, M.; Söding, J.; Lupas, A.N.; Alva, V. Protein sequence analysis using the MPI bioinformatics toolkit. *Curr. Protoc. Bioinform.* **2020**, *72*, e108. [CrossRef]
41. Holm, L.; Sander, C. The FSSP database: Fold classification based on structure-structure alignment of proteins. *Nucleic Acids Res.* **1996**, *24*, 206–209. [CrossRef]
42. Ravantti, J.; Bamford, D.; Stuart, D.I. Automatic comparison and classification of protein structures. *J. Struct. Biol.* **2013**, *183*, 47–56. [CrossRef] [PubMed]
43. Holm, L. DALI and the persistence of protein shape. *Protein Sci.* **2020**, *29*, 128–140. [CrossRef] [PubMed]
44. Friedland, G.D.; Kortemme, T. Designing ensembles in conformational and sequence space to characterize and engineer proteins. *Curr. Opin. Struct. Biol.* **2010**, *20*, 377–384. [CrossRef] [PubMed]
45. Campbell, E.; Kaltenbach, M.; Correy, G.J.; Carr, P.D.; Porebski, B.T.; Livingstone, E.K.; Afriat-Jurnou, L.; Buckle, A.M.; Weik, M.; Hollfelder, F.; et al. The role of protein dynamics in the evolution of new enzyme function. *Nat. Chem. Biol.* **2016**, *12*, 944–950. [CrossRef]
46. Neri, M.; Anselmi, C.; Cascella, M.; Maritan, A.; Carloni, P. Coarse-Grained Model of Proteins Incorporating Atomistic Detail of the Active Site. *Phys. Rev. Lett.* **2005**, *95*, 218102. [CrossRef]
47. Tarenzi, T.; Calandrini, V.; Potestio, R.; Carloni, P. Open-Boundary Molecular Mechanics/Coarse-Grained Framework for Simulations of Low-Resolution G-Protein-Coupled Receptor–Ligand Complexes. *J. Chem. Theory Comput.* **2019**, *15*, 2101–2109. [CrossRef]
48. Fogarty, A.C.; Potestio, R.; Kremer, K. A multi-resolution model to capture both global fluctuations of an enzyme and molecular recognition in the ligand-binding site. *Proteins Struct. Funct. Bioinform.* **2016**, *84*, 1902–1913. [CrossRef]

49. Fiorentini, R.; Kremer, K.; Potestio, R. Ligand-protein interactions in lysozyme investigated through a dual-resolution model. *Proteins Struct. Funct. Bioinform.* **2020**, *88*, 1351–1360. [CrossRef]
50. Giulini, M.; Rigoli, M.; Mattiotti, G.; Menichetti, R.; Tarenzi, T.; Fiorentini, R.; Potestio, R. From system modeling to system analysis: The impact of resolution level and resolution distribution in the computer-aided investigation of biomolecules. *Front. Mol. Biosci.* **2021**, *8*, 676976. [CrossRef]
51. Potestio, R.; Aleksiev, T.; Pontiggia, F.; Cozzini, S.; Micheletti, C. ALADYN: A web server for aligning proteins by matching their large-scale motion. *Nucleic Acids Res.* **2010**, *38*, W41–W45. [CrossRef]
52. Defays, D. An efficient algorithm for a complete link method. *Comput. J.* **1977**, *20*, 364–366. [CrossRef]
53. Marsili, M.; Mastromatteo, I.; Roudi, Y. On sampling and modeling complex systems. *J. Stat. Mech. Theory Exp.* **2013**, *2013*, P09003. [CrossRef]
54. Cubero, R.J.; Jo, J.; Marsili, M.; Roudi, Y.; Song, J. Statistical criticality arises in most informative representations. *J. Stat. Mech. Theory Exp.* **2019**, *2019*, 063402. [CrossRef]
55. Marsili, M.; Roudi, Y. Quantifying relevance in learning and inference. *Phys. Rep.* **2022**, *963*, 1–43. [CrossRef]
56. Mele, M.; Covino, R.; Potestio, R. Information-theoretical measures identify accurate low-resolution representations of protein configurational space. *arXiv* **2022**, arXiv:2205.08437.
57. Holtzman, R.; Giulini, M.; Potestio, R. Making sense of complex systems through resolution, relevance, and mapping entropy. *arXiv* **2022**, arXiv:2203.00100.
58. Cheney, W.; Kincaid, D. Linear algebra: Theory and applications. *Aust. Math. Soc.* **2009**, *110*, 544–550.
59. Schymkowitz, J.; Borg, J.; Stricher, F.; Nys, R.; Rousseau, F.; Serrano, L. The FoldX web server: An online force field. *Nucleic Acids Res.* **2005**, *33*, W382–W388. [CrossRef]
60. Fiser, A.; Do, R.K.G.; Šali, A. Modeling of loops in protein structures. *Protein Sci.* **2000**, *9*, 1753–1773. [CrossRef]
61. Webb, B.; Sali, A. Comparative protein structure modeling using MODELLER. *Curr. Protoc. Bioinform.* **2016**, *54*, 5–6. [CrossRef]
62. Abraham, M.J.; Murtola, T.; Schulz, R.; Páll, S.; Smith, J.C.; Hess, B.; Lindahl, E. GROMACS: High performance molecular simulations through multi-level parallelism from laptops to supercomputers. *SoftwareX* **2015**, *1*, 19–25. [CrossRef]
63. Lindorff-Larsen, K.; Piana, S.; Palmo, K.; Maragakis, P.; Klepeis, J.L.; Dror, R.O.; Shaw, D.E. Improved side-chain torsion potentials for the Amber ff99SB protein force field. *Proteins Struct. Funct. Bioinform.* **2010**, *78*, 1950–1958. [CrossRef] [PubMed]
64. Jorgensen, W.L.; Chandrasekhar, J.; Madura, J.D.; Impey, R.W.; Klein, M.L. Comparison of simple potential functions for simulating liquid water. *J. Chem. Phys.* **1983**, *79*, 926–935. [CrossRef]
65. Bussi, G.; Donadio, D.; Parrinello, M. Canonical sampling through velocity rescaling. *J. Chem. Phys.* **2007**, *126*, 014101. [CrossRef] [PubMed]
66. Parrinello, M.; Rahman, A. Polymorphic transitions in single crystals: A new molecular dynamics method. *J. Appl. Phys.* **1981**, *52*, 7182–7190. [CrossRef]
67. Hess, B.; Bekker, H.; Berendsen, H.J.; Fraaije, J.G. LINCS: A linear constraint solver for molecular simulations. *J. Comput. Chem.* **1997**, *18*, 1463–1472. [CrossRef]
68. Darden, T.; York, D.; Pedersen, L. Particle mesh Ewald: An N log (N) method for Ewald sums in large systems. *J. Chem. Phys.* **1993**, *98*, 10089–10092. [CrossRef]
69. McGibbon, R.T.; Beauchamp, K.A.; Harrigan, M.P.; Klein, C.; Swails, J.M.; Hernández, C.X.; Schwantes, C.R.; Wang, L.P.; Lane, T.J.; Pande, V.S. MDTraj: A modern open library for the analysis of molecular dynamics trajectories. *Biophys. J.* **2015**, *109*, 1528–1532. [CrossRef]
70. Humphrey, W.; Dalke, A.; Schulten, K. VMD: Visual molecular dynamics. *J. Mol. Graph.* **1996**, *14*, 33–38. [CrossRef]
71. Neurath, H.; Walsh, K.A.; Winter, W.P. Evolution of Structure and Function of Proteases: Amino acid sequences of proteolytic enzymes reflect phylogenetic relationships. *Science* **1967**, *158*, 1638–1644. [CrossRef]
72. López-Otín, C.; Bond, J.S. Proteases: Multifunctional enzymes in life and disease. *J. Biol. Chem.* **2008**, *283*, 30433–30437. [CrossRef] [PubMed]
73. Hedstrom, L. Serine protease mechanism and specificity. *Chem. Rev.* **2002**, *102*, 4501–4524. [CrossRef] [PubMed]
74. Verma, S.; Dixit, R.; Pandey, K.C. Cysteine proteases: Modes of activation and future prospects as pharmacological targets. *Front. Pharmacol.* **2016**, *7*, 107. [CrossRef]
75. Rawlings, N.D.; Tolle, D.P.; Barrett, A.J. MEROPS: The peptidase database. *Nucleic Acids Res.* **2004**, *32*, D160–D164. [CrossRef]
76. Rawlings, N.D.; Barrett, A.J.; Finn, R. Twenty years of the MEROPS database of proteolytic enzymes, their substrates and inhibitors. *Nucleic Acids Res.* **2016**, *44*, D343–D350. [CrossRef] [PubMed]
77. Maguid, S.; Fernandez-Alberti, S.; Ferrelli, L.; Echave, J. Exploring the common dynamics of homologous proteins. Application to the globin family. *Biophys. J.* **2005**, *89*, 3–13. [CrossRef]
78. He, Y.; Maisuradze, G.G.; Yin, Y.; Kachlishvili, K.; Rackovsky, S.; Scheraga, H.A. Sequence-, structure-, and dynamics-based comparisons of structurally homologous CheY-like proteins. *Proc. Natl. Acad. Sci. USA* **2017**, *114*, 1578–1583. [CrossRef]
79. Gayathri, P.; Satheshkumar, P.; Prasad, K.; Nair, S.; Savithri, H.; Murthy, M. Crystal structure of the serine protease domain of Sesbania mosaic virus polyprotein and mutational analysis of residues forming the S1-binding pocket. *Virology* **2006**, *346*, 440–451. [CrossRef]
80. Khan, S.; Mian, H.S.; Sandercock, L.E.; Chirgadze, N.Y.; Pai, E.F. Crystal structure of the passenger domain of the Escherichia coli autotransporter EspP. *J. Mol. Biol.* **2011**, *413*, 985–1000. [CrossRef]

81. Sun, D.; Chen, S.; Cheng, A.; Wang, M. Roles of the picornaviral 3C proteinase in the viral life cycle and host cells. *Viruses* **2016**, *8*, 82. [CrossRef]
82. Choi, H.K.; Lee, S.; Zhang, Y.P.; McKinney, B.R.; Wengler, G.; Rossmann, M.G.; Kuhn, R.J. Structural analysis of Sindbis virus capsid mutants involving assembly and catalysis. *J. Mol. Biol.* **1996**, *262*, 151–167. [CrossRef] [PubMed]
83. David, C.C.; Jacobs, D.J. Characterizing protein motions from structure. *J. Mol. Graph. Model.* **2011**, *31*, 41–56. [CrossRef] [PubMed]
84. Lu, D.; Fütterer, K.; Korolev, S.; Zheng, X.; Tan, K.; Waksman, G.; Sadler, J.E. Crystal structure of enteropeptidase light chain complexed with an analog of the trypsinogen activation peptide. *J. Mol. Biol.* **1999**, *292*, 361–373. [CrossRef]
85. Kishi, T.; Kato, M.; Shimizu, T.; Kato, K.; Matsumoto, K.; Yoshida, S.; Shiosaka, S.; Hakoshima, T. Crystal structure of neuropsin, a hippocampal protease involved in kindling epileptogenesis. *J. Biol. Chem.* **1999**, *274*, 4220–4224. [CrossRef] [PubMed]
86. John, S.E.S.; Tomar, S.; Stauffer, S.R.; Mesecar, A.D. Targeting zoonotic viruses: Structure-based inhibition of the 3C-like protease from bat coronavirus HKU4—The likely reservoir host to the human coronavirus that causes Middle East Respiratory Syndrome (MERS). *Bioorganic Med. Chem.* **2015**, *23*, 6036–6048. [CrossRef] [PubMed]
87. Xu, J.; Hu, S.; Wang, X.; Zhao, Z.; Zhang, X.; Wang, H.; Zhang, D.; Guo, Y. Structure basis for the unique specificity of medaka enteropeptidase light chain. *Protein Cell* **2014**, *5*, 178–181. [CrossRef] [PubMed]
88. Kabsch, W.; Sander, C. Dictionary of protein secondary structure: Pattern recognition of hydrogen-bonded and geometrical features. *Biopolym. Orig. Res. Biomol.* **1983**, *22*, 2577–2637. [CrossRef]
89. Touw, W.G.; Baakman, C.; Black, J.; Te Beek, T.A.; Krieger, E.; Joosten, R.P.; Vriend, G. A series of PDB-related databanks for everyday needs. *Nucleic Acids Res.* **2015**, *43*, D364–D368. [CrossRef]

Article

In-Silico Characterization of von Willebrand Factor Bound to FVIII

Valentina Drago [1,†], Luisa Di Paola [2,†], Claire Lesieur [3], Renato Bernardini [4,5], Claudio Bucolo [5] and Chiara Bianca Maria Platania [1,5,*]

1   Clinical Pharmacology and Toxicology Residency Program, Department of Biomedical and Biotechnological Sciences, University of Catania, Via Santa Sofia 97, 95123 Catania, Italy
2   Unit of Chemical-Physics Fundamentals in Chemical Engineering, Department of Engineering, Università Campus Bio-Medico di Roma, Via Álvaro del Portillo 21, 00128 Rome, Italy
3   University of Lyon, CNRS, INSA Lyon, Université Claude Bernard Lyon 1, Ecole Centrale de Lyon, Ampère, UMR5005, 69622 Villeurbanne, France
4   Unit of Clinical Toxicology, Policlinico "G. Rodolico", University of Catania, Via Santa Sofia 97, 95123 Catania, Italy
5   Department of Biomedical and Biotechnological Sciences, University of Catania, Via Santa Sofia 97, 95123 Catania, Italy
*   Correspondence: chiara.platania@unict.it
†   These authors contributed equally to this work.

Featured Application: The computational approaches hereby shown can be used in the rational design of biologic drugs.

Abstract: Factor VIII belongs to the coagulation cascade and is expressed as a long pre-protein (mature form, 2351 amino acids long). FVIII is deficient or defective in hemophilic A patients, who need to be treated with hemoderivatives or recombinant FVIII substitutes, i.e., biologic drugs. The interaction between FVIII and von Willebrand factor (VWF) influences the pharmacokinetics of FVIII medications. In vivo, full-length FVIII (FL-FVIII) is secreted in a plasma-inactive form, which includes the B domain, which is then proteolyzed by thrombin protease activity, leading to an inactive plasma intermediate. In this work, we analyzed through a computational approach the binding of VWF with two structure models of FVIII (secreted full-length with B domain, and B domain-deleted FVIII). We included in our analysis the atomic model of efanesoctocog alfa, a novel and investigational recombinant FVIII medication, in which the VWF is covalently linked to FVIII. We carried out a structural analysis of VWF/FVIII interfaces by means of protein–protein docking, PISA (Proteins, Interfaces, Structures and Assemblies), and protein contact networks (PCN) analyses. Accordingly, our computational approaches to previously published experimental data demonstrated that the domains A3-C1 of B domain-deleted FVIII (BDD-FVIII) is the preferential binding site for VWF. Overall, our computational approach applied to topological analysis of protein–protein interface can be aimed at the rational design of biologic drugs other than FVIII medications.

Keywords: hemophilia A; FVIII; von Willebrand Factor; protein contact networks; bioinformatics; biologic drugs

Citation: Drago, V.; Di Paola, L.; Lesieur, C.; Bernardini, R.; Bucolo, C.; Platania, C.B.M. In-Silico Characterization of von Willebrand Factor Bound to FVIII. *Appl. Sci.* **2022**, *12*, 7855. https://doi.org/10.3390/app12157855

Academic Editors: Robert Jernigan and Domenico Scaramozzino

Received: 4 July 2022
Accepted: 2 August 2022
Published: 4 August 2022

Publisher's Note: MDPI stays neutral with regard to jurisdictional claims in published maps and institutional affiliations.

1. Introduction

Hemophilic A patients are treated periodically with the coagulation factor FVIII substitutes, such as purified hemoderivatives or FVIII biological drugs, which are currently biological advanced therapies approved for treatment of hemophilia A. This class of drugs includes recombinant full-length FVIII and B domain-deleted FVIII [1]. B domain-deleted FVIII (BDD-FVIII) biologics have been developed to improve the biotechnological production of these proteins. Furthermore, BDD-FVIII conjugated with Fc immunoglobulin fragment was reported to have the highest plasma half-life, providing the opportunity of scheduling a low number of infusions [2]. Therefore, BDD-FVIII biologics (products with extended half-life, EHL) have been claimed to decrease the number of infusions in hemophilic patients. EHL products may allow less frequent dosing; however, due to inter-patient differences in FVIII plasma stability and clearance, less frequent infusions may cause longer time periods with relatively low FVIII plasma levels, which could increase the risk of bleeding. The safety cutoff for plasma FVIII levels was set to >1%, and patients with levels >12% were subjected to less bleeding events, especially at joints [2,3]. Specifically, clinical trials carried out so far highlighted a similar capability of different FVIII substitutes on bleeding prevention, measured as annualized bleeding rate [2]. Interestingly, a pharmacokinetic (PK) modeling study predicted that mean values of FVIII plasma levels were similar in patients treated with either BDD-FVIII or full-length recombinant FVIII [1]. Furthermore, this study predicted that patients treated with full-length FVIII (infusions every 48 h) spent more time with FVIII above the 10 IU dL-1 than patients treated with BDD-FVIII product infusions every 72 h. Therefore, full-length FVIII could be characterized by higher plasma stability than B domain-deleted FVIII substitutes compared to BDD-FVIII [1]. This data raised a controversy [4], but the high plasma stability of FL-FVIII was confirmed at pre-clinical level [5].

FVIII half-life is strongly influenced by von Willebrand Factor (VWF), which is reported to bind with FVIII [6–8]. Several studies reported that the binding sites of VWF at FVIII are in the A3 and C1 FVIII domains [9–12] Therefore, with the aim to improve BDD-FVIII plasma stability, a novel FVIII substitute has been developed, efanesoctocog alfa (BIVV001). Efanesoctocog alfa is an investigational biologic drug where VWF D′-D3 domain is covalently bound (through a cleavable XTEN polypeptide linker) to an engineered BDD-FVIII [13]. The atomic model of cryo-electron microscopy (Cryo-EM) of efanesoctocog alfa was built and deposited on Protein Data Bank (PDB:7KWO); in this atomic model VWF binding to A3 and C1 domains of FVIII is confirmed. Mature FVIII is a 2351 a.a pre-protein that, during processing in the endoplasmic reticulum, is cleaved and reassembled in a heavy and a light chain, interacting by means of Van der Waals interactions, and through a metal complex interaction with a divalent cation (Figure 1).

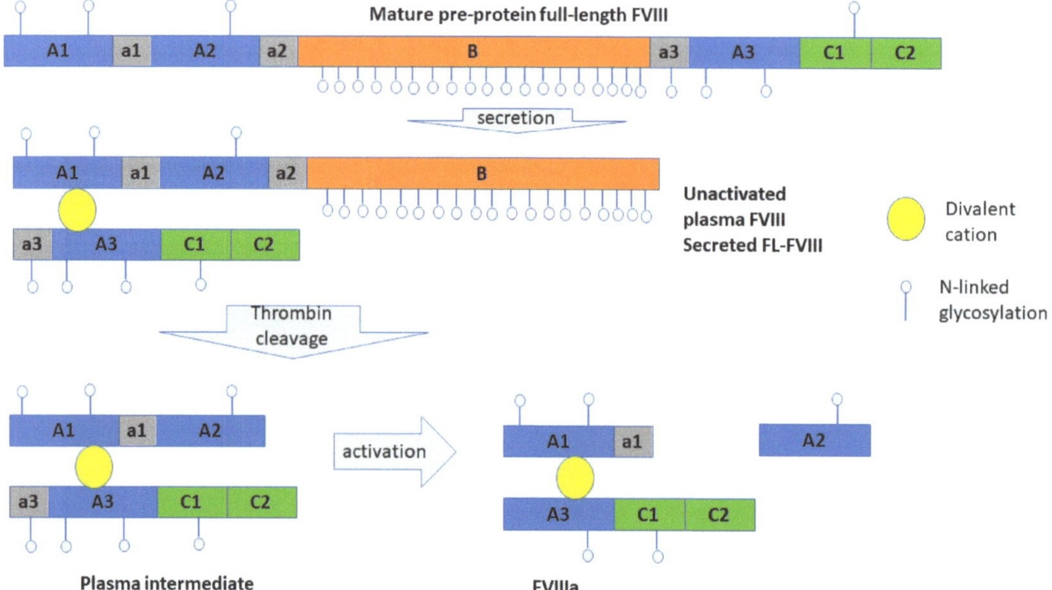

Figure 1. FVIII processing. Capital letters refer to protein domains, while lowercase letters refer to loops linking two protein domains (modified from Pipe SW. Haemophilia 2009 [14]).

After secretion, the processed full-length FVIII (FL-FVIII) is cleaved mainly at a.a 1313, corresponding to the secreted inactivated form of FVIII, containing about 572 amino acids of B domain (secreted full-length FVIII) (Figure 1) [14,15]. However, several heterogenous forms of secreted FL-FVIII were found in plasma [16]. The domain B function has been linked to protection of FVIII against premature proteolysis. Moreover, B domain can inhibit FVIII binding to activated platelets, decreasing overall the inactivation rate of FVIII. B domain has also been involved in modulation of FVIII clearance through binding to the asialoglycoprotein receptor [14]. Finally, FVIII stability to aggregation events is driven by the content of B domain, since FVIII aggregation rate increases with shortening of B domain length [16]. Protein aggregation should generally be avoided for biologics, due to the increased risk of immunogenicity, adverse drug reactions, and modification of pharmacokinetics and pharmacodynamic properties, which then affect overall drug efficacy [17–19]. FVIII medication immunogenicity has been accounted to formation of anti-FVIII antibodies in patients (i.e., inhibitor formation); however, to date, no significant differences in FVIII inhibitor formation have been found in patients treated with full length or BDD-FVIII [20]. Since the stoichiometry of VWF binding to FVIII has not been univocally defined [21–24], and the VWF increases FL-FVIII stability [5,20], we explored through structural computational approach the binding of VWF to modeled B domain of FVIII. To date, the structure of B domain of FVIII has not been solved, although this domain has been identified in magnified Cryo-EM images [25]. Moreover, we reported a structural analysis of protein–protein interactions through molecular docking and structural analysis of protein—protein interfaces [26,27]. We also applied the Protein Contact Networks (PCN) methodology to analyze the topology of the VWF/FVIII complexes, which was validated on more than 1000 protein systems [28]. This methodology, applied to the prediction of interface binding energy, can depict the structure–function relationship in protein–protein complexes along with identification of allosteric binding sites [29–33].

Recently, this approach has also been applied to the analysis of the molecular mechanism behind the SARS-CoV2 infection, analyzing the protein–protein interactions of spike protein/ACE2 complex, providing insight also in the development of new therapeutic strategies [34–37]. Specifically, in our study we included computed novel PCN descriptors of protein–protein interfaces, which identified the key residues involved in the protein–protein interactions of FVIII/VWF complexes.

2. Materials and Methods

Structures have been retrieved from the Protein DataBank as PDB files: 2R7E (B domain-deleted FVIII), 6N29 (D′D3 von Willebrand factor binding domain to FVIII), 7KWO (D′D3 VWF bound to B-domainless FVIII, atomic model of efanesoctocog alfa).

2.1. Structure Modeling

The structure of secreted full-length FVIII was predicted through a two-step modeling approach: (i) the B domain was modeled with the I-Tasser web server; (ii) full-length FVIII model was built with the Advanced Molecular Modeling task of Schrodinger Maestro, by sequence alignment of available structure of B domain-deleted FVIII and modeled B domain, using as input primary sequences of heavy and light chains as reported in Figure 1. Five models of B domain structure were generated with I-Tasser. These models of B domain were used to build 5 models of secreted full-length FVIII, with the Advanced Molecular Modeling task of Schrodinger Maestro. However, 4 models of B domain were automatically excluded by the Advanced Molecular Modeling task due to steric clashes between other protein domains, and just 1 model (Figure S1, secondary structure plot–Supplementary Materials) was further optimized with automated energy minimization steps in the Advanced Molecular Modeling task of Schrodinger Maestro. After energy minimization steps, the optimized models of full-length FVIII (secreted FL-FVIII) and B domain-deleted FVIII (BDD-FVIII) were then subjected to protein–protein docking with a the D′D3 domain of von Willebrand Factor (PDB: 6N29) through PyDock, which provided the prediction of binding free energy of protein complexes (https://life.bsc.es/pid/pydock/ from 2016 to 2022) [38]. Pydock output also includes the scoring of predicted complexes. Rescoring of predicted complexes was also carried out with Prodigy web server (http://milou.science.uu.nl/services/PRODIGY/ from 2016 to 2022) [39].

2.2. Protein Contact Networks

We built the protein contact networks (PCNs) on FVIII complexes starting from the .pdb files, as previously described [27]. In a PCN, protein residues are the network nodes. Links between nodes are active contacts between residues, e.g., when the inter-residue distance lies between 4 and 8 Å, to account for non-covalent residue–residue interactions.

The mathematical representation of the PCN is given by the adjacency matrix, defined as:

$$A_{ij} = \begin{cases} 1 & if\ 4 < d_{ij} < 8 \\ 0 & otherwise \end{cases} \tag{1}$$

where d_{ij} is the Euclidean distance between the i-th and j-th residue, defined as:

$$d_{ij} = \sqrt{\left(x_i - x_j\right)^2 + \left(y_i - y_j\right)^2 + \left(z_i - z_j\right)^2} \tag{2}$$

where $P_i = \{x_i,\ y_i,\ z_i\}$ and $P_j = \{x_j,\ y_j,\ z_j\}$ are the coordinates in a cartesian space of the i-th and j-th residues, respectively (represented by the coordinates of their α-carbons).

Once the PCN was built, we computed the node degree k_i for the i-th node, defined as the number of its links with other nodes, computed as the sum of the elements on the i-th row of the adjacency matrix A:

$$k_i = \sum_j A_{ij} \tag{3}$$

In order to characterize the topology of the protein–protein interactions, for each protein–protein interface we identified links between nodes belonging to the different interfacing chains and, accordingly, we introduced the inter-chain degree of each node k_i^{IC} as the number of links it shares with residues belonging different protein chains.

Nodes (residues) endowed with high inter-chain degree are defined as network hotspots of the protein complex interface, addressing their significant role in protein–protein interactions.

The energy of a graph E is defined as the sum of the absolute values of the adjacency matrix A eigenvalues. Although this is a purely topological descriptor, it captures some physical energy properties of the protein molecular structures [40], particularly oligomers interactions [41].

Focusing on a given interface between two chains, A_i and A_j, the overall inter-chain degree for a given interface $\sum k_{A_iA_j}$ is computed as the sum of the inter-chain degree of residues belonging to a single chain, characterizing the overall interface strength. We defined the average inter-chain degree as the average inter-chain degree value over the number of residues participating in the interface.

We adopted the geometrical descriptors of protein interfaces according to the method of Mei et al. in [42] for each interface between two chains in the complex: 1. the total number of residues Q for each chain in the interface; this number is in general lower than the total interface degree, due to multiple links between residues participating to the interface; 2. the length of the chain involved in the interface R; 3. the interface "roughness" Q/R (previously introduced [9]); 4. the interface amino acid range, $IAR = R/N$ being N the total number of residues in the chain.

For a given interface between two chains, A_i and A_j, the average value of the inter-chain degree is simply given as:

$$< k_{A_iA_j} > = \frac{\sum k_{A_iA_j}}{Q_{A_i} + Q_{A_j}} \tag{4}$$

We introduced energy descriptors, including the topological description provided by the PCNs method. Considering that the interaction energy is higher if the contact distance is smaller, we introduced a weight for each contact:

$$e_{ij} = \frac{1}{d_{ij}} \tag{5}$$

which is also the generic element of the interface energy matrix E, defined as:

$$E = E_{ij} = \begin{cases} e_{ij} = \frac{1}{d_{ij}} & \textit{if } 4 < d_{ij} < 8 \textit{ and the residues belong to different chains} \\ 0 & \textit{otherwise} \end{cases} \tag{6}$$

For each interface, we sorted out the corresponding minor of the interface energy matrix E (corresponding as indices to the rectangular minors of the adjacency matrix) and we introduced the overall interface energy E_{INT} as the sum of e_{ij} for each of the active links at the interface, and the average value $\langle E_{INT} \rangle$ over the whole number of residues at the interface. We also analyzed the single residue contribution to the interface energy, defining the:

$$k_i^{INT} = \sum_j E_{ij} \tag{7}$$

The overall value of energy of interaction $\sum k_{A_i A_j}^{EM}$ is then computed as the sum of all contributions given by Equation (7) for all contacts between chain A_i and A_j. The average value of the interaction energy is given by:

$$< k_{A_i A_j}^{EM} > = \frac{\sum k_{A_i A_j}^{EM}}{Q_{A_i} + Q_{A_j}} \tag{8}$$

Finally, we can define the graph energy of the interface $E_{A_i A_j}$ as the difference between the graph energy of the complex minus the graph energy computed for the single chains (that is, considering the eigenvalues of the adjacency matrix minors corresponding to the single chains).

Furthermore, we completed the analysis through a thermodynamic analysis of the protein complexes via the PISA web server [8] (https://www.ebi.ac.uk/pdbe/pisa/ from 2016 to 2022), reporting from the analysis the following properties: for monomers, a. number of residues exposed at the surface; b. solvent-accessible surface area (ASA) in Å^2; c. solvation free energy of folding of the corresponding structures ΔG_{SOLV} in kcal/mol; for interfaces, a. number of residues exposed at the interface (not accessible to solvent); b. interface area in Å^2 for each monomer (surface area, accessible to solvent in the monomer and no more accessible upon interface formation); c. solvation free energy gain upon formation of the interface $\Delta\Delta G_{SOLV}$ in kcal/mol; the value was calculated as difference in the total solvation energies of isolated and interfacing structures; negative $\Delta\Delta G_{SOLV}$ corresponds to hydrophobic interfaces, or positive protein affinity.

2.3. Network Clustering and Participation Coefficient Calculations

Finally, we applied a network spectral clustering algorithm to identify functional domains in all different conformations [43]; the methodology is based on the spectral decomposition of the network Laplacian, defined as:

$$L = D - A \tag{9}$$

where D is the degree matrix, a diagonal matrix whose diagonal is the degree vector, and A is the network adjacency matrix, as defined in Equation (1). Cluster partition is based on the value of the Fiedler vector v_2 (the eigenvector corresponding to the second minor eigenvalue of L): the cluster number n_c is user-defined. The v_2 components interval

$r_2 = \{\min(v_2), \max(v_2)\}$ is divided into n_c subintervals, so that nodes (residues) are parted in clusters according to which subinterval their v_2 components fall into.

On the basis of network clustering, the participation coefficient P is defined as:

$$P_i = 1 - \left(\frac{k_{si}}{k_i}\right)^2 \tag{10}$$

where k_{si} is the number of links the i-th node shares with nodes belonging to its own cluster.

The participation coefficient is able to identify residues' role in transmitting signals between functional protein regions (protein network clusters) [30,34,40,44].

The PCN methodology is now implemented in open-source software [45].

We projected values of participation coefficient as b-factor and colored the ribbon structures of the complexes by means of an in-house Python script, according to the method previously described [27].

3. Results

3.1. Protein Docking

Protein–protein docking studies were carried out with PyDock on the model of full-length secreted form of FVIII, to predict VWF/FL-FVIII complex (Figure 2). To validate the protein–protein docking and the computational structure interface analysis approaches, we also docked the BDD-FVIII (PDB: 2R7E) with the fragment of von Willebrand factor (PDB: 6N29) (Figure 3). Pydock predicted that von Willebrand factor interacts with the B domain of secreted full-length FVIII with slightly more negative predicted binding free energy, compared to the BDD-FVIII/VWF complex (Table 1).

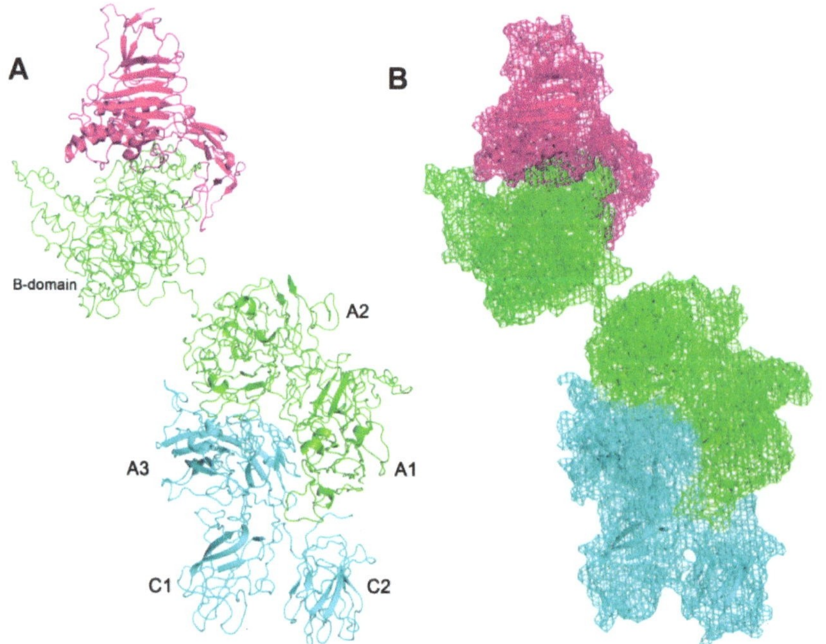

Figure 2. VWF/FL-FVIII complex. (**A**) FL-FVIII is represented in cyan (light-chain) and green cartoons (heavy-chain, including domain (**B**)), VWF is represented with magenta cartoon. (**B**) Surfaces representation as mesh. Capital letters refer to protein domains of FVIII.

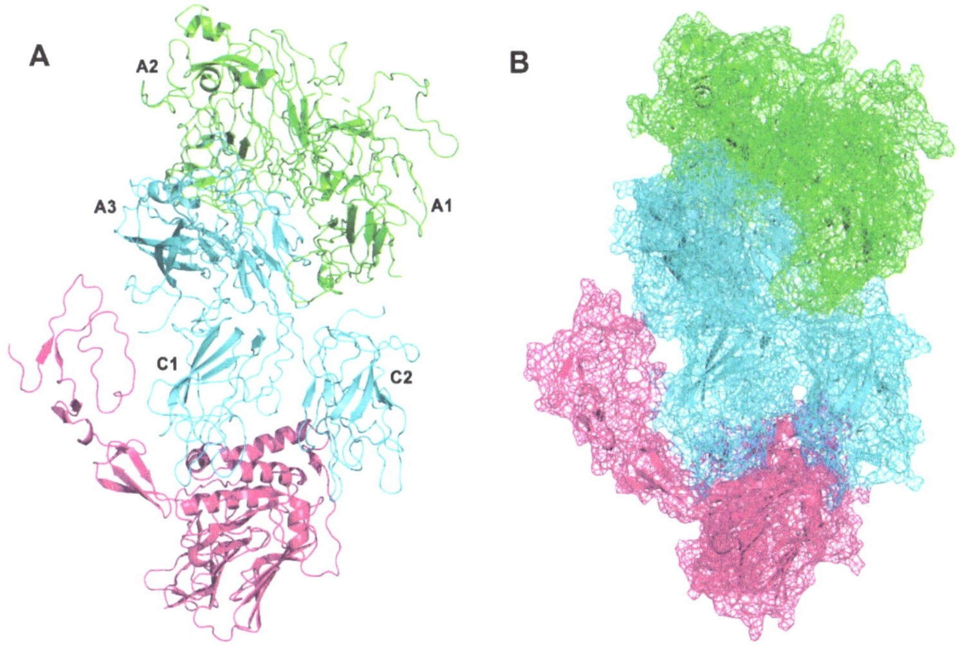

Figure 3. VWF/BDD-FVIII complex. (**A**) BDD-FVIII is represented in cyan (light-chain) and green cartoon, VWF is represented with magenta cartoon. (**B**) Surfaces representation as mesh. Capital letters refer to protein domains of FVIII.

Table 1. Pydock output. Predicted interactions between FVIII structure models and von Willebrand factor fragment (PyDock).

Complex	FVIII-Interacting Domain	Electrostatic Component	Desolvation Component	VDW Component	$\Delta G_{binding}$
B domain-deleted FVIII	A3-C1-light chain	−44.864	−8.012	71.540	−45.722
Secreted full-length FVIII	B domain	−37.144	−10.343	−6.838	−48.172

Shiltagh et al. in 2014 identified the D′D3 (PDB: 6N29) as the domain of von Willebrand factor (VWF) that interacts with FVIII [46]. Chiu et al. in 2015 found that the VWF interacts with the A3 and C1 domain in light chain of the B domain-deleted FVIII [47]. The Pydock output for VWF/BDD-FVIII complex is in accordance with the study results from Chiu et al. 2015 and Fuller et al. 2021 (Figures 3 and 4) [13,47].

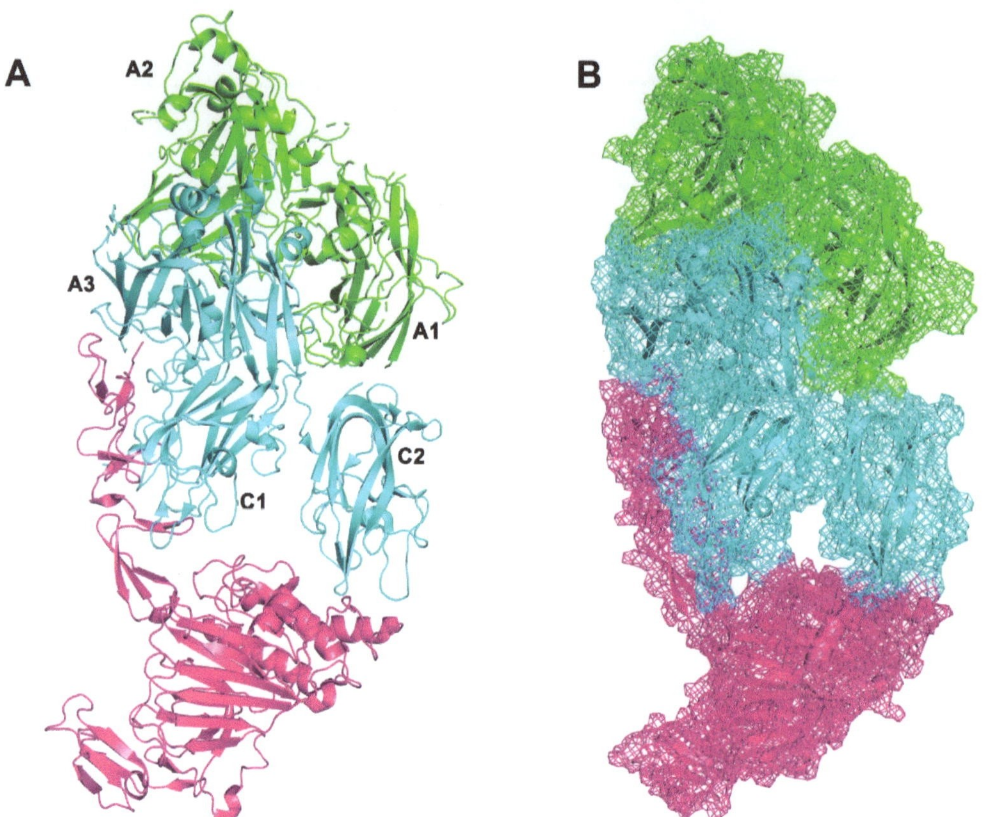

Figure 4. Efanesoctocog alfa (**A**) FVIII is represented in green cartoon (heavy chain) and cyan cartoon (light chain). VWF is represented with magenta cartoon. (**B**) Surfaces representation as mesh. Capital letters refer to protein domains of FVIII.

Fuller et al., 2021, have solved the structure of the bioengineered clinical-stage FVIII substitute, BIVV001 or efanesoctocog alfa (PDB: 7KWO). In efanesoctocog alfa, the VWF-D′D3 is covalently linked to an Fc domain of a BDD-FVIII, resulting in a stabilized VWF-D′D3/BDD-FVIII complex (Figure 4).

The two structures of predicted VWF/BDD-FVIII and of efanesoctocog alfa have been superimposed, and RMSD of alignment was 1.4 Å. PRODIGY rescoring (Table 2) provided binding energy and dissociation constants for analyzed complexes. The binding energy (-14.1 kcal/mol, pKd 10.4) of the predicted complex VWF/FL-FVIII, characterized by the binding of VWF with B domain, was higher (less favorable), although comparable to the binding energy of VWF bound to the A3-C1 domains of BDD-FVIII (-15.3 kcal/mol, pKd 11.3), resembling the slight differences predicted by PyDock. We included the prediction of binding free energy of VWF in efanesoctocog alfa. In the efanesoctocog alfa atomic model, the predicted binding free energy of protein–protein interactions between VWF and FVIII was more negative (~4 logs), compared to binding free energy of the other complexes predicted through protein–protein docking (VWF/FL-FVIII and VWF/BDD-FVIII) (Table 2). Therefore, our computational approach is in accordance with the experimental findings reporting highly stable VWF and FVIII interactions in efanesoctocog alfa or BIVV001.

Table 2. PRODIGY rescoring of PYDOCK-predicted complexes.

Complex	FVIII-Interacting Domains	$\Delta G_{binding}$ (kcal/mol)	pKd
BDD-FVIII	A3-C1 light chain	-15.3	11.3
Efanesoctocog alfa (BIVV001)–atomic model (PDB: 7KWO.)	A3-C1 light chain n	-19.5	14.3
Secreted full-length FVIII	B domain–heavy chain	-14.1	10.4

3.2. Interface Analysis

Since binding free energy prediction with PyDock and PRODIGY, generally, provides an indication of binding driving forces, we further focused our calculations on specific protein–protein interface descriptors. We analyzed the protein–protein interfaces of VWF/FVIII complexes: B domain-deleted FVIII (Table 3), secreted full-length FVIII (Table 4), and efanesoctocog alfa (Table 5). Two interfaces were analyzed: (i) one that involves the heavy and the light chains interactions; (ii) and the VWF/FVIII interface. BDD-FVIII and the FL-FVIII showed similar heavy/light chain interfaces (Tables 3 and 4) in terms of $\Delta\Delta G_{SOLV}$, which was slightly lower in the VWF/FL-FVIII (more favorable) compared to BDD-FVIII complex. These results are coherent with PyDock and Prodigy predictions. Furthermore, according to prediction of binding free energy (PRODIGY calculations) the interactions in the VWF/FVIIII interface of the efanesoctocog alfa were characterized by the most favorable (lowest) predicted free energy $\Delta\Delta$Gsolv, kcal/mol (Table 5), compared to the other analyzed complexes. This lower $\Delta\Delta$Gsolv, kcal/mol (more favorable) corresponded to higher $\Delta\Delta$Gsolv, kcal/mol at interface between heavy and light chains in efanesoctocog alfa, compared to other analyzed complexes. Therefore, it is likely that more stable interactions between VWF/FVIII correspond to a destabilization of FVIII heavy/light chain interface. The results reported in Tables 3–5 indicated that the efanesoctocog alfa is the most stable VWF/FVIII complex, in terms of specific stability of monomers and interface interaction, compared to the VWF/BDD-FVIII and VWF/FL-FVIII complexes, according to findings of Fuller et al. [13].

Table 3. PISA results for the VWF/BDD-FVIII complex. Values within brackets refer to the value specific per residue. Accessible surface area (ASA) of interface (Å^2).

	Protein Residues	Interface Residues	ASA, Å^2	ΔG_{solv}, kcal/mol (Per Residue)
MON H (heavy chain)	693	681	38,466.4	-608.4 (-0.88)
MON L (light chain)	644	624	34,017.4	-570.1 (-0.89)
MON V (ligand VWF)	428	406	23,642.1	-370.7 (-0.87)
INTERFACE H-L				
INTERACTING RESIDUES, MON H		90		
INTERACTING RESIDUES, MON L		100		
INTERFACE AREA, Å^2 (per interacting residue)		3322.2 (17.50)		
$\Delta\Delta$Gsolv, kcal/mol (per interacting residue)		-42.7 (-0.22)		
INTERFACE L-V				
INTERACTING RESIDUES, MON L		42		
INTERACTING RESIDUES, MON V		57		
INTERFACE AREA, Å^2 (per interacting residue)		1742.1 (17.60)		
$\Delta\Delta$Gsolv, kcal/mol (per interacting residue)		-13.3 (-0.13)		

Table 4. PISA results for the VWF/FL-FVIII complex. Values within brackets refer to the value specific per residue. Accessible surface area (ASA) of interface (Å^2).

	Protein Residues	Interface Residues	ASA, Å^2	ΔG_{solv}, kcal/mol (Per Residue)
MON H (heavy chain)	1312	1283	66,894.3	$-881.3\ (-0.67)$
MON L (light chain)	644	627	33,987.6	$-567.8\ (-0.88)$
MON V (ligand VWF)	428	407	23,645.2	$-370.7\ (-0.87)$
INTERFACE H-L				
INTERACTING RESIDUES, MON H				83
INTERACTING RESIDUES, MON L				91
INTERFACE AREA, Å^2 (per interacting residue)				3039 (17.5)
$\Delta\Delta$Gsolv, kcal/mol (per interacting residue)				$-43.8\ (-0.25)$
INTERFACE H (B domain)-V				
INTERACTING RESIDUES, MON H				47
INTERACTING RESIDUES, MON V				49
INTERFACE AREA, Å^2 (per interacting residue)				1496.3 (15.59)
$\Delta\Delta$Gsolv, kcal/mol (per interacting residue)				$-14.3\ (-0.15)$

Table 5. PISA results for efanesoctocog alfa. Values within brackets refer to the value specific per residue. Accessible surface area (ASA) of interface (Å^2).

	Protein Residues	Interface Residues	ASA, Å^2	ΔG_{solv}, kcal/mol (Per Residue)
MON H (heavy chain)	585	522	25,123.4	$-610.6\ (-1.04)$
MON L (light chain)	615	548	28,830.8	$-636.5\ (-1.04)$
MON C (ligand VWF)	478	461	26,423.4	$-567.8\ (-1.19)$
INTERFACE H-L				
INTERACTING RESIDUES, MON H				90
INTERACTING RESIDUES, MON L				82
INTERFACE AREA, Å^2 (per interacting residue)				3049.4 (17.7)
$\Delta\Delta$Gsolv, kcal/mol (per interacting residue)				$-29.0\ (-0.17)$
INTERFACE L-V				
INTERACTING RESIDUES, MON L				67
INTERACTING RESIDUES, MON V				72
INTERFACE AREA, Å^2 (per interacting residue)				2443.6 (17.58)
$\Delta\Delta$Gsolv, kcal/mol (per interacting residue)				$-26.8\ (-0.19)$

To obtain further insight in the protein–protein interactions, we carried out PCN analysis of the three complexes: VWF/BDD-FVIII, efanesoctocog alfa, and VWF/FL-FVIII. Results are shown in Table 6, according to the description provided in Materials and Methods.

PCN analysis (Table 6) has shown that the interface roughness, (Q/R) was very similar in analyzed FVIII structure models. The highest Q/R value was associated to heavy chain of FL-FVIII, because FL-FVIII includes the aminoacids of B domain. In fact, the differences in IAR can be accounted to different length of protein chains in contact with the ligand, the VWF.

Table 6. Topological descriptors for the complex interfaces. Q is the total number of residues for each chain in the interface. R is the length of the chain involved in the interface. Q/R is the interface "roughness." *IAR = R/N* is the interface amino acid range, where N is the total number of residues in the chain. $E_{A_iA_j}$ is the graph energy of the interface. $\sum k_{A_iA_j}$ is the inter-chain degree. $<k_{A_iA_j}>$ is the average value of the inter-chain degree. $\sum k_{A_iA_j}^{EM}$ is the energy of interface interaction, computed as the sum of all contributions for all contacts between chain A_i and A_j. $<k_{A_iA_j}^{EM}>$ is the average value of the interaction energy. H is the label for heavy chain of FVIII, L is the label for light chain of FVIII, and V is the label for VWF. In the VWF/FL-FVIII complex, the VWF interacts with B domain in the heavy chain of FVIII.

	Q_{A_i}	$(\frac{Q}{R})_{A_i}$	IAR_{A_i}	$E_{A_iA_j}$	$\sum k_{A_iA_j}$	$<k_{A_iA_j}>$	$\sum k_{A_iA_j}^{EM}$	$\langle k_{A_iA_j}^{EM}\rangle$
			efanesoctocog alfa (PDB: 7WKO)					
H	63	0.13	0.85	42.86	144	1.17	21.10	0.17
L	60	0.11	0.88					
L	44	0.08	0.87	31.20	95	1.07	14.29	0.16
V	45	0.14	0.67					
			VWF/FL-FVIII					
L	662	0.37	0.94	81.07	4442	3.41	21.1	0.18
H	641	1.00	1.00					
H	31	0.13	0.38	22.66	75	1.15	8.97	0.20
V	34	0.11	0.72					
			VWF/BDD-FVIII					
H	54	0.10	0.82	40.49	146	1.27	21.42	0.19
L	61	0.11	0.84					
L	33	0.14	0.36	28.92	94	1.32	14.31	0.20
V	38	0.12	0.75					

The PCN analysis, as regards the topology of heavy/light chain interface ($H{:}L$), shows that the number of residues involved into active links between heavy and light chains was higher for the FL-FVIII, compared to the BDD-FVIII and the atomic model of efanesoctocog alfa. This result diverges from PISA output, but PCN and PISA reside on two different methods. However, the "absolute interface energy" $E_{A_iA_j}$ at heavy/light ($H{:}L$) chain interface was higher (more favorable) in the VWF/FL-FVIII (81.07 a.u.), compared to values of the VWF/BDD-FVIII (40.49 a.u.) and efanesoctocog alfa (42.89 a.u.) complexes. This trend of energy values, calculated with PCN, are in accordance with interface energy values calculated with PISA. Moreover, the "absolute interface energy" $E_{A_iA_j}$ is proportional to the average inter-chain degree, which was greater at $H{:}L$ interface of FL-FVIII, compared to other VWF/FVIII complexes.

Looking at FVIII/VWF interface, the values of "absolute interface energy" $E_{A_iA_j}$ were higher (more favorable) for the efanesoctocog alfa (31.20 a.u.) and BDD-FVIII (28.92 a.u.), compared to FL-FVIII (22.66 a.u.); indeed, PCN parameters have been in accordance with experimental data and PISA calculations. These differences were mirrored by other topological PCN parameters, i.e., $\sum k_{A_iA_j}^{EM}$ the average value of the interaction energy.

Particularly in the comparison between efanesoctocog alfa and BDD-FVIII, we found subtle differences in graph energy parameters regarding topology of protein–protein interfaces. Therefore, we carried out PCN clustering and participation coefficient (P) calculation, in order to identify through a quantitative approach (see method section, i.e., 2.3 Network clustering and participation coefficient calculations) residues involved in allosteric modulation of protein structure, i.e., allosteric residues (Figures 5–7). The VWF/FL-FVIII (around 1%, Figure 5) complexes showed the lowest number of allosteric residues, compared to the VWF/BDD-FVIII (about 4%, Figure 6) and the efanesoctocog alfa (more than 5%, Figure 7).

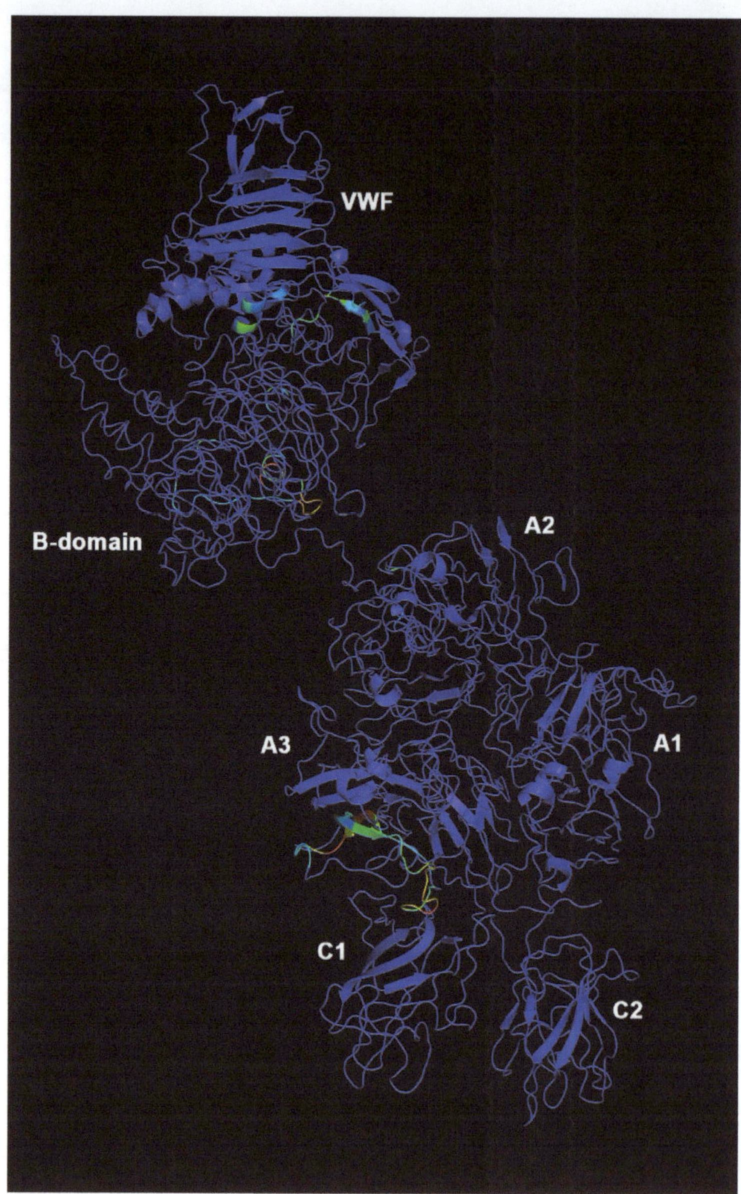

Figure 5. VWF/FL-FVIII complex allosteric residues. Participation coefficient *P* heat map in color scale (blue to red, increasing values of *P*). Blue residues have a *P* = 0. Capital letters refer to protein domains of FVIII.

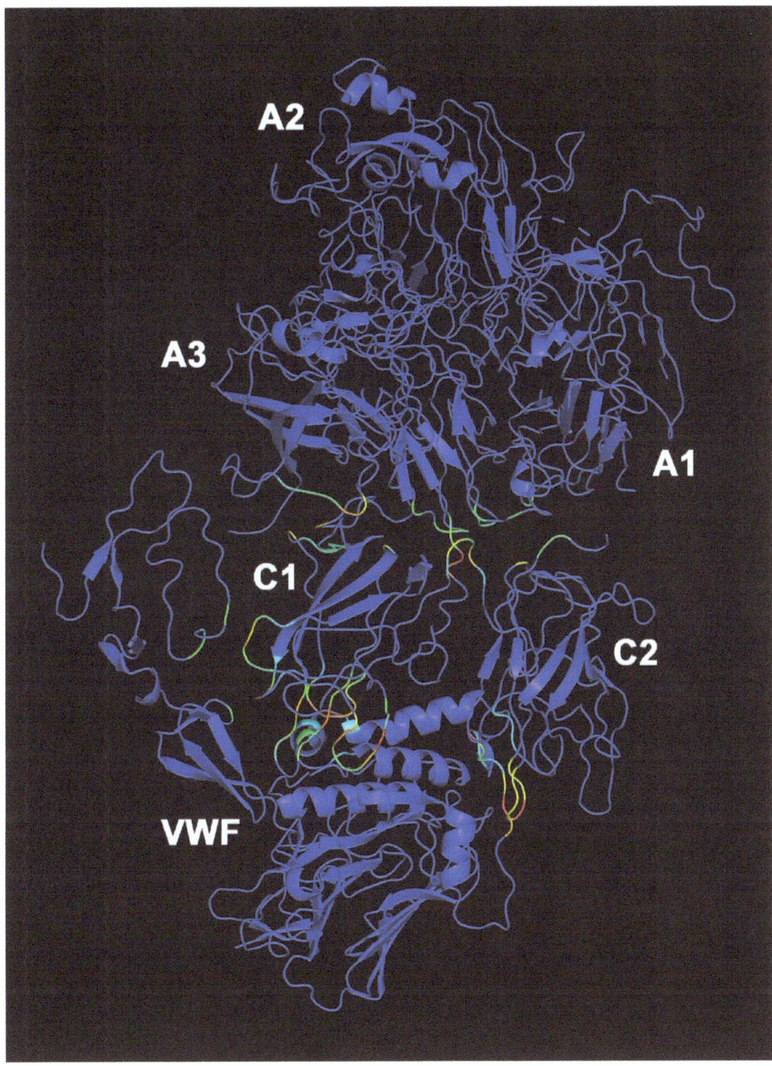

Figure 6. VWF/BDD-FVIII complex allosteric residues. Participation coefficient *P* heat map in color scale (blue to red, increasing values of *P*). Blue residues have a *P* = 0. Capital letters refer to protein domains of FVIII.

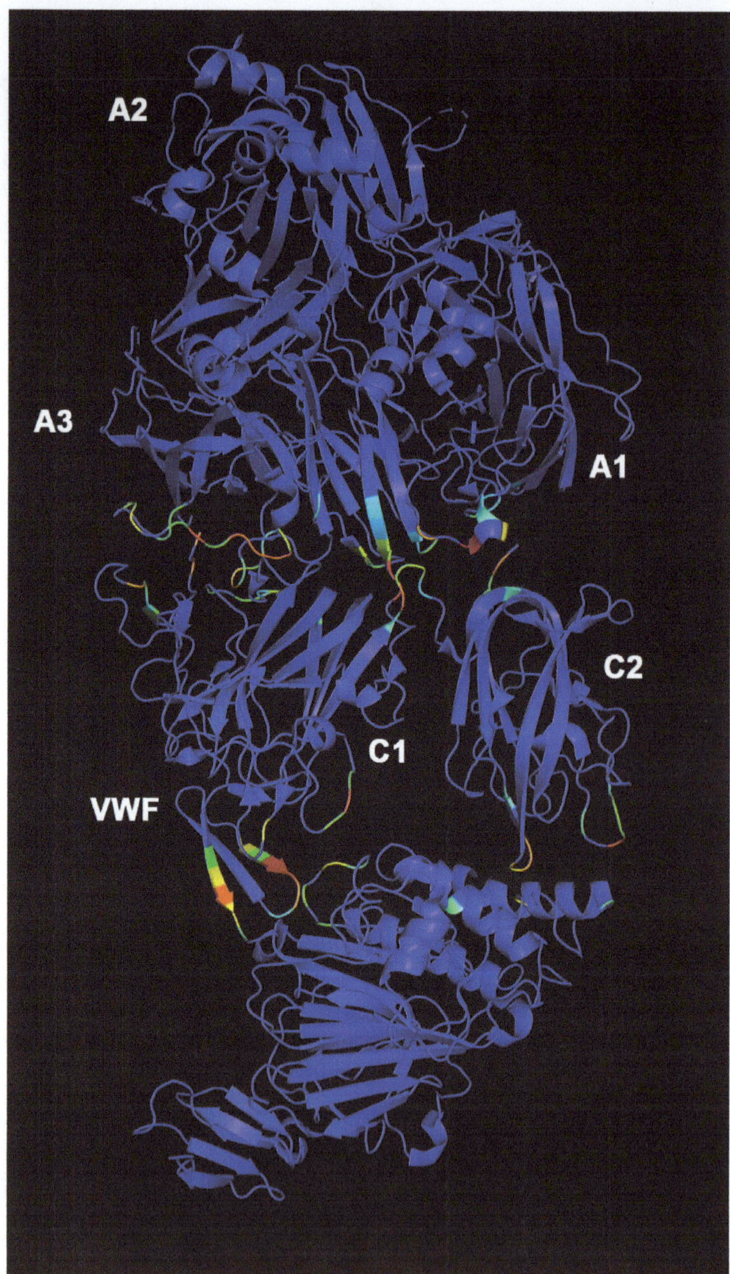

Figure 7. Efanesoctocog alfa allosteric residues. Participation coefficient *P* heat map in color scale (blue to red, increasing values of *P*). Blue residues have a *P* = 0. Capital letters refer to protein domains of FVIII.

The few allosteric residues (high *P* values) in the FL-FVIII/VWF complex (Figure 5) are localized at domain-B/VWF interface, domain B, and A3-C1 interface. The main difference between VWF/BDD-FVIII (Figure 6) and efanesoctocog alfa (Figure 7) stands in the localization and number of allosteric residues. Particularly in VWF/BDD-FVIII

(Figure 6), allosteric residues ($P > 0$) are more distributed in the C1 domain of heavy chain at interface with VWF, compared to efanesoctocog alfa (Figure 7), where most of the allosteric residues (high coefficient P value) are located at A3-C1 and A3-C2 domain interfaces.

These results support the validity and accordance of different computational methods hereby applied: protein–protein docking, docking rescoring, PISA, and PCN analysis of protein and protein domains interfaces.

4. Discussion

FVIII is a key protein involved in the coagulation cascade, and genetic defects in the FVIII gene (F8) cause hemophilia A, an x-linked recessive inherited disease. Hemophilia A is a rare, life-threatening disease affecting 1 in 6000 males, which causes spontaneous and prolonged hemorrhages due to FVIII deficiency. FVIII replacement therapy is the major therapeutic strategy for treatment of hemophilia A, and FVIII medications are listed by the World Health Organization (WHO) as essential medicines [48]. FVIII substitutes consist of full-length FVIII extracted and purified from human plasma, along with several biologic drugs, such as recombinant full-length FVIII (rFL-FVIII) and recombinant B domain-deleted FVIII (rBDD-FVIII). BDD-FVIII products have been developed to improve production yield and standardize recombinant protein production processes, but rBDD-FVIII were also found to be associated to increased FVIII plasma half-life; therefore, these products are also denominated as extended half-life (EHL) FVIII [49]. Interaction between von Willebrand Factor (VWF) and FVIII has been reported to improve FVIII plasma stability and pharmacokinetics properties, without modification of FVIII pharmacodynamics [8,9,50]. From this perspective, a novel FVIII investigational replacement medication has been developed, i.e., the BIVV001 or efanesoctocog alfa, which is a recombinant fusion protein in which the B domain-deleted FVIII is covalently linked to VWF, through a XTEN polypeptide linker [13]. VWF binds with most favorable energy to A3-C1 domains of FVIII light chain, however, binding of VWF to other domains of FVIII should not be excluded, since FVIII and VWF stoichiometry has not been univocally identified [21,22]. Moreover, the VWF binding to full-length FVIII has been linked to increased FVIII stability and decreased FVIII immunogenicity (i.e., inhibitor formation). This explains the trend of FL-FVIII medication to show a longer half-life over time and a similar or lower rate of inhibitor formation in treatment-naïve patients, compared to BDD-FVIII medications [5,8,20]. These data can also be attributed to binding of B domain to other plasma proteins, such as albumin [20]. Our in silico study could be considered as a small step to translational investigation, because it outlined how B domain in FL-FVIII/VWF would be an additional, but not the most favorable, binding site for VWF, thus putatively contributing to high plasma stability of FL-FVIII medication.

Our in silico study provided a structural insight on the binding of von Willebrand Factor to FVIII, and our computational data are in accordance with the experimental findings, i.e., FVIII A3-C1 domain is the most stable binding site of VWF. Our computational approach suggested that one of the driving forces of VWF binding at this preferential binding site could be related to conformational modifications of FVIII, through modulation of allosteric residues.

The PISA and PCN analysis of topological parameters suggested an unfavorable binding of VWF at B domain of FVIII, due to an increased stability of the interface between the heavy and the light chain, compared to VWF/BDD-FVIII and efanesoctocog alfa. In fact, the most stable VWF/FVIII complex (efanesoctocog alfa) was characterized by highest (less favorable) interaction energy between heavy and light chains of FVIII. Differences between the modeled VWF/BDD-FVIII complex and the atomic model of efanesoctocog alfa are not likely to be attributed to the lower resolution of BDD-FVIII X-ray structure (PDB:2R7E), compared to the atomic model of efanesoctocog alfa (PDB: 7KWO). Furthermore, the X-ray structure of BDD-FVIII showed 4% of Phi and Psi angles in disallowed regions of the Ramachandran plot, but these residues are located in disordered loops of A1 and A2 domains of heavy chain, a region not involved in the binding with VWF [51].

Moreover, the PCN clustering and participation coefficient P calculations revealed that VWF binding to the B domain of FL-FVIII was characterized by lower number of residues with participation coefficient $P > 0$ (allosteric residues), compared to VWF/BDD-FVIII and efanesoctocog alfa. The participation coefficient is an output parameter of PCN analysis, and has been found useful for identification of allosteric residues in different protein system and protein domains, according to the methodological approach hereby used and previously applied [31,40]. The topological parameters, coming from PCN analysis, also provided information on the local contribution to interface energy, which is useful for identification of key residues (e.g., allosteric amino acids) in protein–protein complex formation. Therefore, we can hypothesize that one of the driving forces in VWF binding to FVIII is attributed to allosteric modulation of protein structure.

Further investigations can shed light on putative allosteric and cooperative protein–protein interaction, e.g., by simulating FVIII structures with several replicas of VWF. These studies may provide new hints about the structural role of different domains of FVIII, along with interaction with other plasma proteins, such as the serum albumin. One of the main limitations of our study is attributed to the intrinsic approximation structural modeling of B domain, whose structure has never been solved, with the exclusion of Cryo-EM density images [25]. Another limitation is related to approximation of intrinsic and essential FVIII post-translational modifications, such as protein glycosylation (Figure 1), and the most glycosylated domain of FVIII is the B domain.

5. Conclusions

In conclusion, we hereby carried out an integrated computational approach which provided outputs that are in accordance with experimental data: i.e., most favorable binding site for VWF in the FVIII (A3-C1 domains in the light chain). Our computational approach provided new hints about the involvement of domain B of FVIII as another putative, although less favorable, binding site for VWF. Additionally, we can hypothesize, given the accordance between different computational methods, that the most stable VWF/FVIII complex (efanesoctocog alfa) is characterized by most unfavorable interface energy between heavy and light chains of FVIII, paralleled by the most favorable VWF/FVIII interface, likely due to the involvement of the highest number of residues with high participation coefficient (i.e., allosteric residues). Overall, our computational approaches provided new hints on interdomain allosteric communication in proteins or protein–protein complexes, which are considered as one of the driving forces in the protein–protein binding stability. Thereby, our integrated computational approach will be helpful in the rational structure design of biologic drugs.

Supplementary Materials: The following supporting information can be downloaded at: https://www.mdpi.com/article/10.3390/app12157855/s1, Figure S1: Secondary structure plot of secreted full-length FVIII.

Author Contributions: Conceptualization, C.B.M.P. and V.D.; methodology, C.B.M.P. and L.D.P.; software, C.B.M.P., L.D.P. and C.B.; formal analysis, C.B.M.P.; C.L. and L.D.P.; Investigation, V.D., C.B.M.P., C.L. and L.D.P.; resources, C.B. and V.D.; data curation, V.D., C.B.M.P. and R.B.; writing—original, draft preparation, C.B.M.P.; writing—review and editing, C.B.M.P., L.D.P., C.L., V.D., R.B. and C.B.; visualization, C.B.M.P.; supervision, C.B.M.P. and V.D.; project administration, C.B.M.P. and V.D.; funding acquisition, C.B.M.P. and C.B. All authors have read and agreed to the published version of the manuscript.

Funding: C.B.M.P. has been supported by the funding PON Ricerca e Innovazione D.M. 1062/21–Contratti di ricerca, from the Italian Ministry of University (MUR). Contract #: 08-I-17629-2.

Institutional Review Board Statement: Not applicable.

Informed Consent Statement: Not applicable.

Data Availability Statement: Structural models will be provided by the corresponding author upon request of the readers.

Acknowledgments: The manuscript is completed in memory of Salvatore Asero (BIOVIIIx), who contributed to conceptualization of the project. Authors wish to thank the company BIOVIIIx (Via Alessandro Manzoni 1, 80128 Napoli, Italy) for unrestricted support and fruitful scientific discussion. The manuscript is dedicated to Salvatore (Totò) Salomone, who spent his career in the field of cardiovascular pharmacology, and unfortunately departed too early from the Pharmacology section of the BIOMETEC department-University of Catania.

Conflicts of Interest: The authors declare no conflict of interest.

References

1. Gringeri, A.; Wolfsegger, M.; Steinitz, K.N.; Reininger, A.J. Recombinant full-length factor VIII (FVIII) and extended half-life FVIII products in prophylaxis-new insight provided by pharmacokinetic modelling. *Haemophilia* **2015**, *21*, 300–306. [CrossRef] [PubMed]
2. Lieuw, K. Many factor VIII products available in the treatment of hemophilia a: An embarrassment of riches? *J. Blood Med.* **2017**, *8*, 67–73. [CrossRef] [PubMed]
3. Den Uijl, I.E.M.; Fischer, K.; Van Der Bom, J.G.; Grobbee, D.E.; Rosendaal, F.R.; Plug, I. Analysis of low frequency bleeding data: The association of joint bleeds according to baseline FVIII activity levels. *Haemophilia* **2011**, *17*, 41–44. [CrossRef] [PubMed]
4. Shapiro, A.D.; Li, S. Response to Gringeri et al.: "Recombinant full-length factor VIII (FVIII) and extended half-life FVIII products in prophylaxis–new insight provided by pharmacokinetic modelling". *Haemophilia* **2015**, *21*, e489–e492. [CrossRef]
5. Bloem, E.; Karpf, D.M.; Nørby, P.L.; Johansen, P.B.; Loftager, M.; Rahbek-Nielsen, H.; Petersen, H.H.; Blouse, G.E.; Thim, L.; Kjalke, M.; et al. Factor VIII with a 237 amino acid B-domain has an extended half-life in F8-knockout mice. *J. Thromb. Haemost.* **2019**, *17*, 350–360. [CrossRef]
6. Pipe, S.W.; Montgomery, R.R.; Pratt, K.P.; Lenting, P.J.; Lillicrap, D. Life in the shadow of a dominant partner: The FVIII-VWF association and its clinical implications for hemophilia A. *Blood* **2016**, *128*, 2007–2016. [CrossRef]
7. Shi, Q.; Kuether, E.L.; Schroeder, J.A.; Perry, C.L.; Fahs, S.A.; Gill, J.C.; Montgomery, R.R. Factor VIII inhibitors: Von Willebrand factor makes a difference in vitro and in vivo. *J. Thromb. Haemost.* **2012**, *10*, 2328–2337. [CrossRef]
8. Vollack-Hesse, N.; Oleshko, O.; Werwitzke, S.; Solecka-Witulska, B.; Kannicht, C.; Tiede, A. Recombinant VWF Fragments Improve Bioavailability of Subcutaneous Factor VIII in Hemophilia A Mice. *Blood* **2020**, *137*, 1072–1081. [CrossRef]
9. Yee, A.; Gildersleeve, R.D.; Gu, S.; Kretz, C.A.; McGee, B.M.; Carr, K.M.; Pipe, S.W.; Ginsburg, D. A von Willebrand factor fragment containing the D′D3 domains is sufficient to stabilize coagulation factor VIII in mice. *Blood* **2014**, *124*, 445–452. [CrossRef]
10. Saenko, E.L.; Scandella, D. The acidic region of the factor VIII light chain and the C2 domain together form the high affinity binding site for von Willebrand factor. *J. Biol. Chem.* **1997**, *272*, 18007–18014. [CrossRef]
11. Dagil, L.; Troelsen, K.S.; Bolt, G.; Thim, L.; Wu, B.; Zhao, X.; Tuddenham, E.G.; Nielsen, T.E.; Tanner, D.A.; Faber, J.H.; et al. Interaction Between the a3 Region of Factor VIII and the TIL′E′ Domains of the von Willebrand Factor. *Biophys. J.* **2019**, *117*, 479–489. [CrossRef] [PubMed]
12. Gilbert, G.E.; Kaufman, R.J.; Arena, A.A.; Miao, H.; Pipe, S.W. Four hydrophobic amino acids of the factor VIII C2 domain are constituents of both the membrane-binding and von Willebrand factor-binding motifs. *J. Biol. Chem.* **2002**, *277*, 6374–6381. [CrossRef] [PubMed]
13. Fuller, J.R.; Knockenhauer, K.E.; Leksa, N.C.; Peters, R.T.; Batchelor, J.D. Molecular determinants of the factor VIII/von Willebrand factor complex revealed by BIVV001 cryo-electron microscopy. *Blood* **2021**, *137*, 2970–2980. [CrossRef]
14. Pipe, S.W. Functional roles of the factor VIII B domain. *Haemophilia* **2009**, *15*, 1187–1196. [CrossRef] [PubMed]
15. Jankowski, M.A.; Patel, H.; Rouse, J.C.; Marzilli, L.A.; Weston, S.B.; Sharpe, P.J. Defining "full-length" recombinant factor VIII: A comparative structural analysis. *Haemophilia* **2007**, *13*, 30–37. [CrossRef]
16. Anzengruber, J.; Feichtinger, M.; Bärnthaler, P.; Haider, N.; Ilas, J.; Pruckner, N.; Benamara, K.; Scheiflinger, F.; Reipert, B.M.; Malisauskas, M. How Full-Length FVIII Benefits from Its Heterogeneity–Insights into the Role of the B-Domain. *Pharm. Res.* **2019**, *36*, 77. [CrossRef]
17. Moussa, E.M.; Panchal, J.P.; Moorthy, B.S.; Blum, J.S.; Joubert, M.K.; Narhi, L.O.; Topp, E.M. Immunogenicity of Therapeutic Protein Aggregates. *J. Pharm. Sci.* **2016**, *105*, 417–430. [CrossRef]
18. Roberts, C.J. Therapeutic protein aggregation: Mechanisms, design, and control. *Trends Biotechnol.* **2014**, *32*, 372–380. [CrossRef]
19. Joubert, M.K.; Luo, Q.; Nashed-Samuel, Y.; Wypych, J.; Narhi, L.O. Classification and characterization of therapeutic antibody aggregates. *J. Biol. Chem.* **2011**, *286*, 25118–25133. [CrossRef]
20. Schiavoni, M.; Napolitano, M.; Giuffrida, G.; Coluccia, A.; Siragusa, S.; Calafiore, V.; Lassandro, G.; Giordano, P. Status of Recombinant Factor VIII Concentrate Treatment for Hemophilia a in Italy: Characteristics and Clinical Benefits. *Front. Med.* **2019**, *6*, 261. [CrossRef]
21. Vlot, A.; Koppelman, S.; Berg, M.V.D.; Bouma, B.; Sixma, J. The affinity and stoichiometry of binding of human factor VIII to von Willebrand factor. *Blood* **1995**, *85*, 3150–3157. [CrossRef] [PubMed]

22. Lollar, P.; Parker, C.G. Stoichiometry of the porcine factor VIII-von Willebrand factor association. *J. Biol. Chem.* **1987**, *262*, 17572–17576. [CrossRef]

23. Fischer, B.E.; Kramer, G.; Mitterer, A.; Grillberger, L.; Reiter, M.; Mundt, W.; Dorner, F.; Eibl, J. Effect of multimerization of human and recombinant von Willebrand factor on platelet aggregation, binding to collagen and binding of coagulation factor VIII. *Thromb. Res.* **1996**, *84*, 55–66. [CrossRef]

24. Fischer, B.E.; Schlokat, U.; Reiter, M.; Mundt, W.; Dorner, F. Biochemical and functional characterization of recombinant von Willebrand factor produced on a large scale. *Cell. Mol. Life Sci.* **1997**, *53*, 943–950. [CrossRef]

25. Grushin, K.; Miller, J.; Dalm, D.; Parker, E.T.; Healey, J.F.; Lollar, P.; Stoilova-McPhie, S. Lack of recombinant factor VIII B-domain induces phospholipid vesicle aggregation: Implications for the immunogenicity of factor VIII. *Haemophilia* **2014**, *20*, 723–731. [CrossRef]

26. Krissinel, E.; Henrick, K. Inference of Macromolecular Assemblies from Crystalline State. *J. Mol. Biol.* **2007**, *372*, 774–797. [CrossRef]

27. Di Paola, L.; Platania, C.B.M.; Oliva, G.; Setola, R.; Pascucci, F.; Giuliani, A. Characterization of Protein-Protein Interfaces through a Protein Contact Network Approach. *Front. Bioeng. Biotechnol.* **2015**, *3*, 170. [CrossRef]

28. Di Paola, L.; Paci, P.; Santoni, D.; De Ruvo, M.; Giuliani, A. Proteins as sponges: A statistical journey along protein structure organization principles. *J. Chem. Inf. Model.* **2012**, *52*, 474–482. [CrossRef]

29. Hu, G.; Di Paola, L.; Liang, Z.; Giuliani, A. Comparative Study of Elastic Network Model and Protein Contact Network for Protein Complexes: The Hemoglobin Case. *BioMed Res. Int.* **2017**, *2017*, 2483264. [CrossRef]

30. Minicozzi, V.; Di Venere, A.; Nicolai, E.; Giuliani, A.; Caccuri, A.M.; Di Paola, L.; Mei, G. Non-symmetrical structural behavior of a symmetric protein: The case of homo-trimeric TRAF2 (tumor necrosis factor-receptor associated factor 2). *J. Biomol. Struct. Dyn.* **2020**, *39*, 319–329. [CrossRef]

31. Platania, C.B.M.; Di Paola, L.; Leggio, G.M.; Romano, G.L.; Drago, F.; Salomone, S.; Bucolo, C. Molecular features of interaction between VEGFA and anti-angiogenic drugs used in retinal diseases: A computational approach. *Front. Pharmacol.* **2015**, *6*, 248. [CrossRef] [PubMed]

32. Platania, C.B.M.; Ronchetti, S.; Riccardi, C.; Migliorati, G.; Marchetti, M.C.; Di Paola, L.; Lazzara, F.; Drago, F.; Salomone, S.; Bucolo, C. Effects of protein-protein interface disruptors at the ligand of the glucocorticoid-induced tumor necrosis factor receptor-related gene (GITR). *Biochem. Pharmacol.* **2020**, *178*, 114110. [CrossRef] [PubMed]

33. Platania, C.B.M.; Bucolo, C. Molecular Dynamics Simulation Techniques as Tools in Drug Discovery and Pharmacology: A Focus on Allosteric Drugs. *Methods Mol. Biol.* **2021**, *2253*, 245–254. [CrossRef] [PubMed]

34. Di Paola, L.; Hadi-Alijanvand, H.; Song, X.; Hu, G.; Giuliani, A. The Discovery of a Putative Allosteric Site in the SARS-CoV-2 Spike Protein Using an Integrated Structural/Dynamic Approach. *J. Proteome Res.* **2020**, *19*, 4576–4586. [CrossRef]

35. Verkhivker, G.M.; Di Paola, L. Integrated Biophysical Modeling of the SARS-CoV-2 Spike Protein Binding and Allosteric Interactions with Antibodies. *J. Phys. Chem. B* **2021**, *125*, 4596–4619. [CrossRef]

36. Verkhivker, G.M.; Di Paola, L. Dynamic Network Modeling of Allosteric Interactions and Communication Pathways in the SARS-CoV-2 Spike Trimer Mutants: Differential Modulation of Conformational Landscapes and Signal Transmission via Cascades of Regulatory Switches. *J. Phys. Chem. B* **2021**, *125*, 850–873. [CrossRef]

37. Hadi-Alijanvand, H.; Di Paola, L.; Hu, G.; Leitner, D.M.; Verkhivker, G.M.; Sun, P.; Poudel, H.; Giuliani, A. Biophysical Insight into the SARS-CoV2 Spike-ACE2 Interaction and Its Modulation by Hepcidin through a Multifaceted Computational Approach. *ACS Omega* **2022**, *7*, 17024–17042. [CrossRef]

38. Rosell, M.; Rodríguez-Lumbreras, L.A.; Romero-Durana, M.; Jiménez-García, B.; Díaz, L.; Fernández-Recio, J. Integrative modeling of protein-protein interactions with pyDock for the new docking challenges. *Proteins Struct. Funct. Bioinform.* **2020**, *88*, 999–1008. [CrossRef]

39. Xue, L.C.; Rodrigues, J.P.; Kastritis, P.L.; Bonvin, A.M.; Vangone, A. PRODIGY: A web server for predicting the binding affinity of protein-protein complexes. *Bioinformatics* **2016**, *32*, 514–3678. [CrossRef]

40. Cimini, S.; Di Paola, L.; Giuliani, A.; Ridolfi, A.; De Gara, L. GH32 family activity: A topological approach through protein contact networks. *Plant Mol. Biol.* **2016**, *92*, 401–410. [CrossRef]

41. Di Paola, L.; Mei, G.; Di Venere, A.; Giuliani, A. Exploring the stability of dimers through protein structure topology. *Curr. Protein Pept. Sci.* **2016**, *17*, 30–36. [CrossRef] [PubMed]

42. Mei, G.; Di Venere, A.; Rosato, N.; Finazzi-Agrò, A. The importance of being dimeric. *FEBS J.* **2005**, *272*, 16–27. [CrossRef] [PubMed]

43. Tasdighian, S.; Di Paola, L.; De Ruvo, M.; Paci, P.; Santoni, D.; Palumbo, P.; Mei, G.; Di Venere, A.; Giuliani, A. Modules identification in protein structures: The topological and geometrical solutions. *J. Chem. Inf. Model.* **2014**, *54*, 159–168. [CrossRef] [PubMed]

44. Di Venere, A.; Nicolai, E.; Minicozzi, V.; Caccuri, A.; Di Paola, L.; Mei, G. The odd faces of oligomers: The case of traf2-c, a trimeric c-terminal domain of tnf receptor-associated factor. *Int. J. Mol. Sci.* **2021**, *22*, 5871. [CrossRef]

45. Guzzi, P.H.; Di Paola, L.; Giuliani, A.; Veltri, P. PCN-Miner: An open-source extensible tool for the Analysis of Protein Contact Networks. *Bioinformatics* **2022**, *7*, btac450. [CrossRef]

46. Shiltagh, N.; Kirkpatrick, J.; Cabrita, L.D.; McKinnon, T.A.J.; Thalassinos, K.; Tuddenham, E.G.D.; Hansen, D.F. Solution structure of the major factor VIII binding region on von Willebrand factor. *Blood* **2014**, *123*, 4143–4151. [CrossRef]

47. Chiu, P.-L.; Bou-Assaf, G.M.; Chhabra, E.S.; Chambers, M.G.; Peters, R.T.; Kulman, J.D.; Walz, T. Mapping the interaction between factor VIII and von Willebrand factor by electron microscopy and mass spectrometry. *Blood* **2015**, *126*, 935–938. [CrossRef]
48. World Health Organization. WHO Model List of Essential Medicines. 2021. Available online: https://www.who.int/publications/i/item/WHO-MHP-HPS-EML-2021.02 (accessed on 1 June 2022).
49. Kessler, C.M.; Gill, J.C.; White, G.C.; Shapiro, A.; Arkin, S.; Roth, D.A.; Meng, X.; Lusher, J.M. B-domain deleted recombinant factor VIII preparations are bioequivalent to a monoclonal antibody purified plasma-derived factor VIII concentrate: A randomized, three-way crossover study. *Haemophilia* **2005**, *11*, 84–91. [CrossRef]
50. Przeradzka, M.A.; Van Galen, J.; Ebberink, E.H.T.M.; Hoogendijk, A.J.; Van Der Zwaan, C.; Mertens, K.; Biggelaar, M.V.D.; Meijer, A.B. D′ domain region Arg782-Cys799 of von Willebrand factor contributes to factor VIII binding. *Haematologica* **2020**, *105*, 1695–1703. [CrossRef]
51. Shen, B.W.; Spiegel, P.C.; Chang, C.-H.; Huh, J.-W.; Lee, J.-S.; Kim, J.; Kim, Y.-H.; Stoddard, B.L. The tertiary structure and domain organization of coagulation factor VIII. *Blood* **2008**, *111*, 1240–1247. [CrossRef]

applied sciences

MDPI

Article

Free-Energy Landscape Analysis of Protein-Ligand Binding: The Case of Human Glutathione Transferase A1

Adrien Nicolaï [1,*], Nicolas Petiot [1], Paul Grassein [1], Patrice Delarue [1], Fabrice Neiers [2] and Patrick Senet [1]

[1]   Laboratoire Interdisciplinaire Carnot de Bourgogne, UMR 6303 CNRS-Université de Bourgogne Franche-Comté, 21078 Dijon, France
[2]   Centre des Sciences du Goût et de l'Alimentation (CSGA), Université de Bourgogne Franche-Comté, INRA, CNRS, 21000 Dijon, France
*   Correspondence: adrien.nicolai@u-bourgogne.fr

Citation: Nicolaï, A.; Petiot, N.; Grassein, P., Delarue, P., Neiers, F., Senet, P. Free-Energy Landscape Analysis of Protein-Ligand Binding: The Case of Human Glutathione Transferase A1. *Appl. Sci.* **2022**, *12*, 8196. https://doi.org/10.3390/app12168196

Academic Editors: Robert Jernigan and Domenico Scaramozzino

Received: 13 July 2022
Accepted: 12 August 2022
Published: 16 August 2022

Publisher's Note: MDPI stays neutral with regard to jurisdictional claims in published maps and institutional affiliations.

Abstract: Glutathione transferases (GSTs) are a superfamily of enzymes which have in common the ability to catalyze the nucleophilic addition of the thiol group of reduced glutathione (GSH) onto electrophilic and hydrophobic substrates. This conjugation reaction, which occurs spontaneously but is dramatically accelerated by the enzyme, protects cells against damages caused by harmful molecules. With some exceptions, GSTs are catalytically active as homodimers, with monomers generally constituted of 200 to 250 residues organized into two subdomains. The first is the N-terminal subdomain, which contains an active site named G site, where GSH is hosted in catalytic conformation and which is generally highly conserved among GSTs. The second subdomain, hydrophobic, which binds the substrate counterpart (H site), can vary from one GST to another, resulting in structures able to recognize different substrates. In the present work, we performed all-atom molecular dynamics simulations in explicit solvent of human GSTA1 in its APO form, bound to GSH ligand and bound to GS-conjugated ligand. From MD, two probes were analyzed to (i) decipher the local conformational changes induced by the presence of the ligand and (ii) map the communication pathways involved in the ligand-binding process. These two local probes are, first, coarse-grained angles (θ, γ), representing the local conformation of the protein main chain and, second, dihedral angles χ representing the local conformation of the amino-acid side chains. From the local probes time series, effective free-energy landscapes along the amino-acid sequence were analyzed and compared between the three different forms of GSTA1. This methodology allowed us to extract a network of 33 key residues, some of them being located in the experimentally well-known binding sites G and H of GSTA1 and others being located as far as 30Å from the original binding sites. Finally, the collective motions associated with the network of key residues were established, showing a strong dynamical coupling between residues Gly14-Arg15 and Gln54-Val55, both in the same binding site (intrasite) but also between binding sites of each monomer (intersites).

Keywords: enzyme; ligand binding; molecular dynamics; free-energy; coarse-grained angles

1. Introduction

Proteins are biological macromolecules that perform a large variety of functions in living cells comprising biochemical (enzymes), structural, mechanical, and signaling functions. To perform their functions, proteins interact with small molecules referred to as ligands, which are able to bind to a protein with high affinity and specificity [1]. These protein/ligand interactions are crucial in biology, particularly in the context of drug design [2]. Since proteins interact with a broad range of drugs, it is of particular interest to study the mechanisms of binding of ligands to proteins and its impact on the structural dynamics to gain insights into (i) phenomena involved in the biological process and related to diseases [3] (misfolding, aggregation), and (ii) discovery, design, and development of new drugs [4]. The experimental structural data (e.g., X-ray crystallography, NMR, or cryo-EM) provide key structural information of the ligand-bound and ligand-unbound (APO) proteins [5]. Nevertheless, the static information is not always sufficient for understanding protein–ligand binding mechanisms, especially when pockets are highly flexible and contain several binding sites. Therefore, molecular dynamics (MD) is a powerful tool that provides a description of the dynamics and structures of protein–ligand systems with a high spatial and temporal resolution.

Glutathione transferases (GSTs) belong to a ubiquitous superfamily of enzymes that metabolize a broad range of reactive toxic compounds by catalyzing the conjugation of reduced tripeptide glutathione (γ-Glu-Cys-Gly; named GSH) to the electrophilic center of a second substrate [6–8], the reactivity of GSH being due to the thiol group SH of the cysteine residue. The conjugation reaction occurs spontaneously but GST accelerates it dramatically. This process of detoxification protects cells against damages caused by both exogenous and endogenous molecules. GSTs were first discovered in liver cells [9,10], and since then, they have been found to exhibit ligand-binding properties for a large variety of compounds, which are not always their enzymatic substrates [11]. Therefore, GSTs participate in diverse biological processes, making them multifunctional proteins. Moreover, GSTs are classified into three families according to their location in the cell: cytosolic, mitochondrial, and microsomal, which is not evolutively related to the two other classes [12]. First-discovered and most-abundant cytosolic GSTs are divided into 13 classes based on homology of their sequences. Members of the same cytosolic class have at least 40% of sequence identity, while members of different classes must have at most 25% of sequence identity. Even if they present a low homology with the cytosolic GST, mitochondrial GSTs can be considered as a particular class of GSTs (Kappa). Humans possess GST members in seven different classes [13], particularly the Alpha-class (cytosolic GSTs), in which the GSTA1 protein, which is the protein of interest in the present work, belongs. An alternative classification is possible on the basis of the residue located in the G site and which favors the activation of GSH (deprotonation of GSH plus reduction of pKa): the Cys-GSTs, whose structures are very similar to the ancestral precursor of all GSTs, the Ser-GSTs (Delta, Theta, Zeta, and Phi classes, including also Nu-GST activated by a threonine), and the Tyr-GSTs (Alpha, Pi, Mu, and Sigma classes); this latter subfamily comprises the more recently evolved GSTs.

With some exceptions, GSTs are catalytically active as dimers. The GSTA1 dimer is stabilized by a "lock and key" motif consisting of two key residues (Met51 and Phe52) fitting into a hydrophobic cavity of the other dimer [14]. Human GSTA1 dimer revealed negative cooperativity properties depending on the substrates. It was proposed that this negative cooperativity allows the self-preservation of their functions [15,16]. Additionally, there is no clear evidence that monomeric forms of GSTA1 are active as well as folded [17]. GST monomers are, in general, made of 200 to 250 residues with a molecular weight generally comprising between 25 and 30 kDa [18,19] and are organized into two subdomains (Figure 1A): the typically 80-residue-long N-terminal subdomain (I) has the typical fold of thioredoxin. It contains a first active site where GSH is hosted in catalytic conformation, named G site (Figure 1B). The thioredoxin fold is composed of a characteristic sequence of α-helices and β-strands encountered in the thioredoxin protein family, i.e., $\beta_1 - \alpha_1 - \beta_2 - \alpha_2 - \beta_3 - \beta_4 - \alpha_3$, which is characteristic of enzymes dealing with gluthatione, such as glutaredoxin or glutathione peroxydase [20]. In addition, it has been shown that the region $\beta_3 - \beta_4 - \alpha_3$ is well conserved among GSTs and enables GSH recognition by the enzyme [7]. Residues forming the G site are generally conserved among GSTs [19]. A noticeable exception is the residue of α_1-helix which interacts with the sulfur atom of GSH and which can be a cysteine, a serine, or a tyrosin (as is the case for human GSTA1). Depending on the nature of this residue, GSTs develop slightly different catalytic functions and target a different range of substrates [21]. The second subdomain (II) is all-helical and contains the H site which binds the substrate. The number of helices varies between four and seven among GSTs. Subdomain (II) is hydrophobic, and therefore attractive for hydrophobic molecules. Together with the G site, the combined architecture of GST monomers is adapted to bind GSH to hydrophobic substrates [22]. Contrary to the G site, the residues of the H site strongly vary from one GST to the others, resulting in H sites of different natures and the ability to recognize different substrates [23].

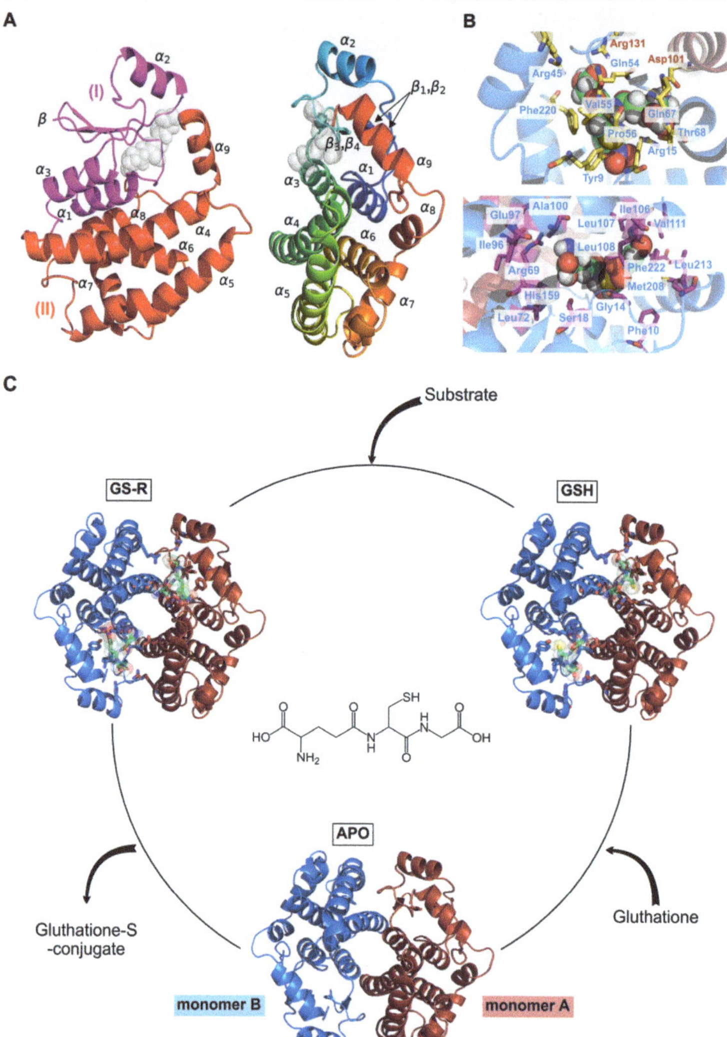

Figure 1. (**A**) Cartoon representation of human GSTA1 monomer structures. Left panel: subdomains (I) and (II) are indicated in magenta and red, respectively. Right panel: the color code is the following: N-term (blue) to C-term (red). Secondary structures labels are also indicated. (**B**) Ligand binding G (top panel) and H (bottom panel) sites of human GSTA1. Ligands are shown in green spheres, and residues belonging to the binding sites are shown in yellow and purple sticks, respectively. The color code is the following: hGSTA1 monomer A in red, hGSTA1 monomer B in blue. (**C**) Catalytic cycle of GSH conjugation to electrophilic substrate. Three forms of hGSTA1 during the conjugation reaction cycle are highlighted: APO (no-compound-bound), GSH (glutathione-bound), and GS-R (gluthatione-S-conjugated substrate). The R form (substrate-bound) is not considered in the present work. The color code is the same as in panel (**B**).

Hereafter, we focus our interest on the structure of human GSTA1 (hGSTA1), which is a homodimer with each monomer made of 222 residues [24]. Among the 29 experimental structures of hGSTA1 available in the Protein Data Bank, 21 of them contain at least one ligand, which is GSH, substrates (R), or GS-conjugates (GS-R). For each structure, a sequence analysis is provided, featuring the residues of hGSTA1 which are in contact with the ligand. Collecting these data (Figure S1), we determined the list of residues bound to GSH and therefore involved in the G site of hGSTA1: Tyr9, Arg15, Arg45, Gln54, Val55, Pro56, Gln67, Thr68, Asp101, Arg131, and Phe220 (Asp101 and Arg131 belong to the opposite monomer; see Figure 1B). We performed the same analysis to identify residues bound to GS-conjugates and therefore involved in the H site of hGSTA1: Phe10, Gly14, Ser18, Arg69, Leu72, Ile96, Glu97, Ala100, Ile106, Leu107, Leu108, Val111, His159, Met208, Leu213, and Phe222. From one experimental X-ray resolved structure, we performed all-atom classical molecular dynamics (MD) in explicit solvent of hGSTA1 enzyme (see Supplementary Materials for details about the MD protocol) in three different ligand-binding forms of the conjugation reaction cycle shown in Figure 1C: (i) its APO form, when there is no compound bound to GST, (ii) its GSH form, when the glutathione is bound to the G-site, and (iii) its GS-R form, when the glutathione-S-conjugated substrate is bound to the G and H sites, the substrate considered here being the 1-chloro-2,4-dinitrobenzene (CDNB). Another existing form of the hGSTA1 enzyme is not considered in the present work, i.e., the R form (substrate-bound). Indeed, the binding of the gluthatione ligand and of the substrate is not sequential. The substrate can bind first to GST and then the GSH, or the opposite way, as presented in Figure 1C.

Overall, the present study aims at deciphering the local conformational changes and shifts of populations of different local minima of the main chain and side chains of hGSTA1 enzyme in its three different ligand-binding forms. Predicting ligand-induced local free-energy changes is relevant both for understanding sequence–structure–function relationships in enzymes and also for structure-based drug design. From MD, we explore local conformational changes using internal coordinates, i.e., coarse-grained angles of the main chain and dihedral angles of the side chains. Dissimilarities between effective free-energy landscapes of the internal coordinates were established in order to identify the network of residues involved in the ligand-binding process of hGSTA1. Finally, the coupling between internal coordinates was revealed by analyzing the correlation of their motions using the principal component analysis, allowing the definition of collective motions to which these degrees of freedom contribute the most.

2. Materials and Methods

2.1. Internal Angles as Local Probes of the Protein Main and Side-Chain Conformational Changes

Local conformational changes of the main chain of human GSTA1 were analyzed using coarse-grained angles (CGAs) (θ, γ), which can be represented by a unit vector in spherical coordinates, where γ is the azimuth angle and θ the polar angle [25]. For a residue i along the amino-acid sequence (see Figure 2A), θ_i is the bond angle formed by the virtual bonds joining three successive C^α atoms ($i - 1$, i and $i + 1$) and γ_i is the dihedral angle formed by the virtual bonds joining four successive C^α atoms ($i - 1$, i, $i + 1$ and $i + 2$). The first pair of CGA (θ, γ) along the sequence is (θ_2, γ_2) and the last one is $(\theta_{N-2}, \gamma_{N-2})$, where N is the total number of residues. The convention for γ angles is the following: each angle varies between $-180°$ and $+180°$, with $\gamma = 0°$ being chosen when C^α_{i-1} is cis to C^α_{i+2} and the clockwise rotation of $C^\alpha_{i+1} - C^\alpha_{i+2}$ is positive when looking from C^α_i to C^α_{i+1}. Because the length of the $C^\alpha - C^\alpha$ virtual bond between two consecutive residues is nearly constant, the main-chain conformation is entirely described by the main-chain bond angles θ and the main-chain torsional angles γ (Figure 2A). These CGAs (θ, γ), which represent the torsion and curvature of the protein main chain and form a complete set of order parameters for protein folding [26], are part of coarse-grained protein models [27] and were used to analyze large conformational changes of proteins [28], protein folding, and dynamics in all-atom simulations [25,29,30] and conformational ensemble of intrinsically disordered proteins [31].

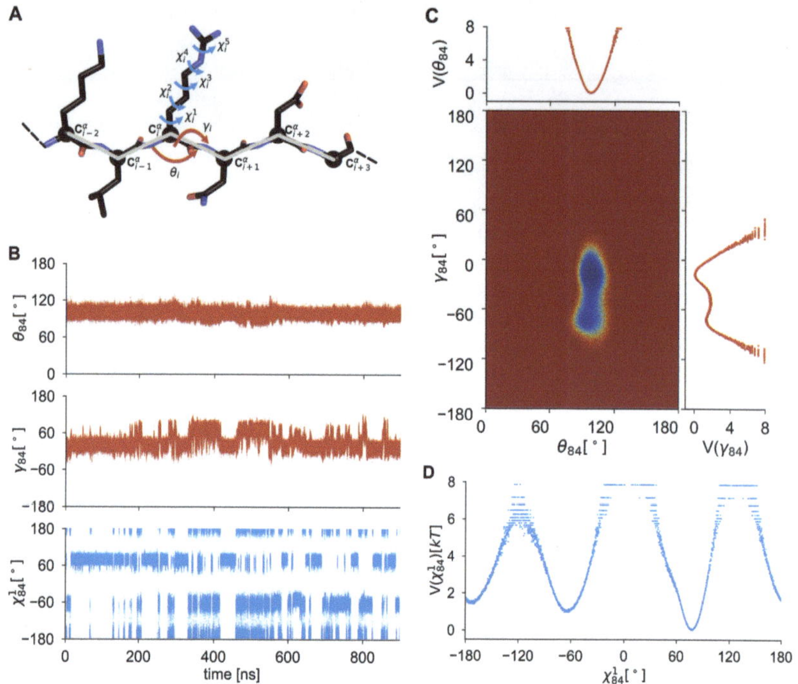

Figure 2. (**A**) Cartoon representation of CGAs (θ, γ) and SCAs (χ^k) used as local probes to track local conformational changes in hGSTA1 from MD. C^α atoms are shown with black spheres. The main chain and side chains of each residue are shown in sticks. (**B**) Time series of CGAs θ (**top panel**), γ (**middle panel**), and SCA χ^1 (**bottom panel**) for residue 84, as an example, recorded during run 1 of hGSTA1 in its APO form. (**C**) Effective free-energy surface $V(\theta, \gamma)_{84}$ computed from time series shown in panel (**B**). Effective free-energy profiles of each internal coordinate θ_{84} and γ_{84} are also presented. (**D**) Effective free-energy profile $V(\chi^1)_{84}$ computed from time series shown in panel B.

Moreover, local conformational changes of the side chains of human GSTA1 were analyzed using side-chain dihedral angles (SCAs) χ^k. SCAs capture the rotation around $C - C$ or $C - N$ bonds of the side chain from its C^β to its extremity. Each SCA is built from the coordinates of four successive atoms along the side chain of an amino acid. First of all, dihedral angle χ^1 is made of $N - C^\alpha - C^\beta - X$, where X depends on the amino acid. X can be S^γ (Cys), O^γ (Ser, Thr), or C^γ for all the other amino acids except Gly and Ala, for which χ^1 dihedral angle is not defined. Second, χ^2 corresponds to the rotation around the bond $C^\beta - C^\gamma$ and is not defined for amino acids Gly, Ala, Cys, Ser, Thr, and Val. Third, χ^3, corresponding to the rotation around the bond $C^\gamma - C^\delta$ or $C^\gamma - S^\delta$ (Met), is only defined for amino acids Gln, Glu, Met, Arg, and Lys. Finally, χ^4 are only defined for Arg and Lys amino acids and correspond to the rotation around the bond $C^\delta - C^\epsilon$ (Lys) or $C^\delta - N^\epsilon$ (Arg), and χ^5 is only defined for Arg and corresponds to the rotation of the side chain around the bond $N^\epsilon - C^\zeta$ (Figure 2A).

2.2. Free-Energy Surface, Free-Energy Profile, and Similarity Index

Effective 2D Free-Energy Surfaces (FESs) $V(\theta_i, \gamma_i)$ were computed for each pair of CGA (θ_i, γ_i) by using

$$V(\theta_i, \gamma_i) = -k_B T \log P(\theta_i, \gamma_i) \tag{1}$$

where k_B is the Boltzmann constant, T is the temperature, and $P(\theta_i, \gamma_i)$ is the probability density function (PDF) of CGA pair (θ_i, γ_i). Two-dimensional PDFs were computed on a sphere, using CGA time series (Figure 2B) extracted from concatenated MD trajectories for each hGSTA1 binding-state: APO (2 runs of 900 ns), GSH (2 runs of 900 ns), and GS-R (1 run of 1200 ns). An example of free-energy surface for CGA pair $(\theta_{84}, \gamma_{84})$ (Gly83-Lys84-Asp85-Ile86) is shown in Figure 2C.

Then, to quantify the modifications between two FESs of identical CGA pair (θ_i, γ_i) due to different ligand-binding form of hGSTA1, i.e., APO vs. GSH, GSH vs. GS-R, or GS-R vs. APO, as shown in Figure 1C, we computed the similarity index H between their associated 2D PDFs using

$$H_{1|2}(\theta_i, \gamma_i) = \frac{2 \int_0^{\pi} \sin\theta_i d\theta_i \int_{-\pi}^{+\pi} P_1(\theta_i, \gamma_i) P_2(\theta_i, \gamma_i) d\gamma_i}{\int_0^{\pi} \sin\theta_i d\theta_i \int_{-\pi}^{+\pi} P_1(\theta_i, \gamma_i)^2 d\gamma_i + \int_0^{\pi} \sin\theta_i d\theta_i \int_{-\pi}^{+\pi} P_2(\theta_i, \gamma_i)^2 d\gamma_i} \tag{2}$$

where $P_1(\theta_i, \gamma_i)$ is the PDF of CGA pair (θ_i, γ_i) in binding form 1 and $P_2(\theta_i, \gamma_i)$ is the PDF of CGA pair (θ_i, γ_i) in binding form 2. The similarity index H varies between 0 (dissimilar) and 1 (identical). We consider similarity between 2 FESs to be (i) large if $H > 0.70$, (ii) moderate if $0.30 \leq H < 0.70$, and (iii) low if $H < 0.3$, as per a previous work [28].

Similarly, effective 1D free-energy profiles (FEPs) $V(\chi_i^k)$ were computed for each SCA χ_i^k ($k = 1, ..., 5$) by using

$$V(\chi_i^k) = -k_B T \log P(\chi_i^k) \tag{3}$$

where $P(\chi_i^k)$ is the PDF of the dihedral angle χ_i^k. One-dimensional PDFs were computed on a circle from time series extracted from MD (Figure 2B) using concatenated MD trajectories, as performed for 2D PDFs (see above). An example of free-energy profile for side-chain dihedral angle χ_{84}^1 (Lys) is shown in Figure 2D. Moreover, modifications between two FEPs of identical side-chain dihedral angle χ_i^k due to different ligand-binding forms of GST were quantified using the similarity index H between their associated 1D PDFs [28]:

$$H_{1|2}(\chi_i^k) = \frac{2 \int_{-\pi}^{+\pi} P_1(\chi_i^k) P_2(\chi_i^k) d\chi_i^k}{\int_{-\pi}^{+\pi} P_1(\chi_i^k)^2 d\chi_i^k + \int_{-\pi}^{+\pi} P_2(\chi_i^k)^2 d\chi_i^k} \tag{4}$$

where $P_1(\chi_i^k)$ is the PDF of SCA (θ_i, γ_i) in binding form 1, and $P_2(\chi_i^k)$ is the PDF of SCA (χ_i^k) in binding form 2. The same scale of similarity, as described above for FESs, is used to quantify similarities (large, moderate, and low).

3. Results

3.1. Structural Flexibility of Human GSTA1 in Its APO, GSH, and GS-R Form

Figure 3A presents thermal B-factors computed during MD simulations of hGSTA1 in its APO, GSH, and GS-R form. First, the correlation between B-factors of monomers A and B in each of the three forms independently is very high (more than 80%). Therefore, we compare average B-factors computed over the two monomers for each form. Overall, four regions of the hGSTA1 enzyme were found to exhibit a large flexibility (Figure 3B): the α_2 region (residues 38–50), which contains residue Arg45 that belongs to the G site; the loop between α_4–α_5 helices (residues 107–119), i.e., $L_{4,5}$, which contains residues Leu107, Leu108, and Val111 that belong to the H site, the loop between α_8–α_9 helices, i.e., $L_{8,9}$ (residues

199–206), and, finally, the C-terminal region (residues 210–222) containing the α_9 helix with residues Phe220 and Phe222 that belong to the G and H sites, respectively. By computing the difference of flexibility of hGSTA1 between the three forms shown in Figure 1C, we found that the APO and GSH forms are more flexible than the GS-R form in the α_2 region, which contains residue Arg45. This flexibility in the absence of the substrate (APO and GSH forms) may promote its binding, the same region being constrained by the presence of the conjugated substrate (GS-R form). On the opposite, the loop and extremities of α_4 and α_5 helices are more flexible in the GS-R and GSH forms of hGSTA1 than in its APO form. It means that the presence of the ligand in the G and H sites of GST and, particularly, the interactions with residues Ala100, Asp101, Ile106, Leu107, Leu108, Val111, (in α_4) and Arg131 (in α_5) modify the flexibility of this region (Figure 1B). Moreover, the presence of GSH in the G site strongly increases the flexibility of the loop $L_{8,9}$ located a few residues away for the C-terminal helix α_9 of hGSTA1. This loop is also slightly flexible in the APO form and is not at all in the GS-R form. Last, but not least, the APO form of hGSTA1 shows a much larger flexibility in the C-terminal region (residue 210–222) than both the GSH and the GS-R forms. The sum of the B-factors in this region for each of the forms is 39, 21, and 12 nm^2 for APO, GSH, and GS-R, respectively. This result is in agreement with experimental observations showing that the protein flexibility of hGSTA1, including the dynamics of C-terminal α_9 helix on nanosecond-millisecond timescale and the protruding extremities of α_4–α_5 helices, contributes remarkably to the catalytic and noncatalytic ligand-binding functions of GSTs [32,33].

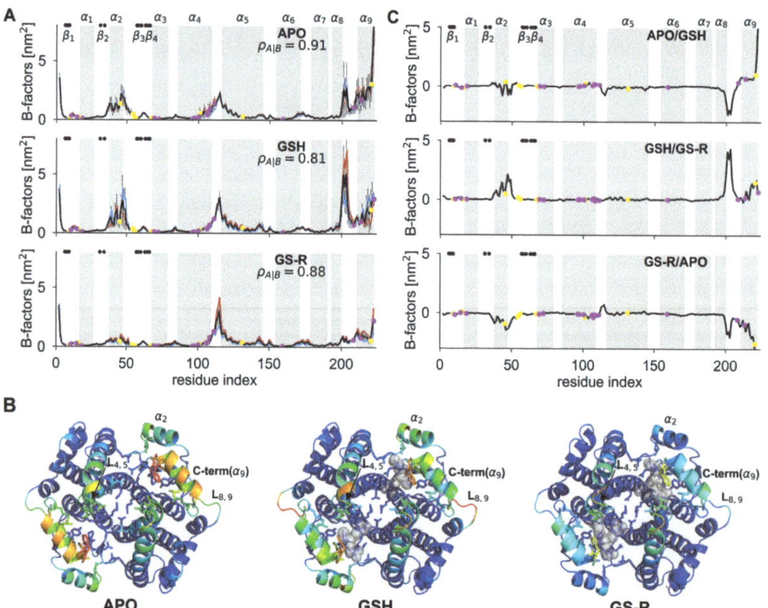

Figure 3. (**A**) Thermal B-factors of hGSTA1 computed from MD in its three different forms: APO (**top**), GSH (**middle**), and GS-R (**bottom**) panel. Yellow and purple circles correspond to residue that belongs to the G and H site, respectively. Gray rectangles indicate residues in α helix, and black circles at the top of each plot represent residues in β-sheets. The Pearson correlation between monomers A and B, $\rho_{A|B}$, is also indicated for each form. (**B**) Cartoon representation of hGSTA1 structures colored according to their B-factors in its three different forms: APO (**top**), GSH (**middle**), and GS-R (**bottom**) panel. Ligand is shown in transparent spheres, and residues which belong to the G and H sites of each monomer are shown in stick. Secondary structures of interest (see text) are also indicated: α_2

helix; the loop between α_4–α_5 helices, i.e., $L_{4,5}$; the loop between α_8–α_9 helices, i.e., $L_{8,9}$ and the C-terminal region (α_9 helix). The color code used is rainbow, from blue to red color corresponding to 0 and 5 nm^2, respectively. (**C**) Difference of B-factors between two different forms of hGSTA1. (**Top**) panel: APO vs. GSH, (**middle**) panel: GSH vs. GS-R, and (**bottom**) panel: GS-R vs. APO.

3.2. Identification of Key Residues Involved in Ligand Binding to hGSTA1 from Free-Energy Landscape Analysis of CGAs and SCAs

The network of residues influenced by ligand binding into hGSTA1 in the three different forms presented in Figure 1C is determined by computing similarity indices H between (i) 2D free-energy surfaces of CGAs (θ, γ) and (ii) 1D free-energy profiles of SCAs (χ) along the amino-acid sequence. First, only FESs that show large or moderate similarities ($H > 0.3$) between monomers A and B of the homodimer are considered as the *DATA*. This ensures that non-converged FESs and FEPs between the two monomers are not taken into account. In fact, even though monomers A and B can have an asymmetrical dynamical behavior from MD due to the fact they are flexible monomers (sampling issue), no information can be exploited from these non-converged characteristics from a thermodynamical point of view. They might or might not be relevant to the conformational fluctuations of the protein. In total, 11 CGAs were excluded ($i = 50$ for APO, $i = 141, 145, 220$ for GSH, and $i = 144, 146, 147, 149, 217, 218, 219, 220$ for GS-R form). These CGAs are mainly located in the loop between helices α_5 and α_6 and in the C-terminal region, and they represent only 5% of the total number of CGAs ($N = 218$).

Second, we computed similarity indices H between FESs of CGAs (from *DATA*) between the different forms of hGSTA1, (i) APO vs. GSH, (ii) GSH vs. GS-R, and (iii) GS-R vs. APO. Figure 4 shows dissimilarity $1 - H$ along the amino-acid sequence for each comparison for both CGAs (θ, γ) (panel A) and SCAs χ^1 (panel B). On one hand, from APO vs. GSH comparison, only one CGA $(\theta, \gamma)_i$ presents moderate similarity of its FES between the two forms. This result is not very surprising since at the microsecond timescale, the association/dissociation of GSH ligands from G sites of monomer A and B of hGSTA1 were observed during MD simulations (see Figure S3). It concerns CGA $(\theta, \gamma)_{143}$ and involves residues Ser142, His143, Gly144, and Gln145, which are not directly involved in the ligand-binding sites of hGSTA1. However, these residues are located at the end of α_5-helix, which contains residue Arg131 of the G site. On the other hand, comparisons involving the GS-R form of hGSTA1 present many more structural modifications. From GSH vs. GS-R comparison, seven CGAs $(\theta, \gamma)_i$ present moderate or low similarities of their FESs between these two forms, for $i = 13, 14, 15, 16, 17, 18, 51$. These seven CGAs involve 13 different amino acids with Gly14, Arg15, and Ser18, which are directly involved in the ligand-binding site of hGSTA1 (Figure 4C). Among the ten other residues detected and which are not located in the binding sites of hGSTA1, Arg20 is of great interest since this amino acid is highly conserved through GST classes. Moreover, by looking at FES of CGA $(\theta, \gamma)_{15}$ in the three ligand-binding forms of hGSTA1 (Figure 4D), we show that both θ and γ angles are modified by the presence of gluathione-S-conjugated ligand compared to the APO and GSH forms. For instance, the global minimum of the FES $V(\theta_{15})$ is significantly shifted to lower values of θ and there is the creation of a new global minimum for $V(\gamma_{15})$ in the GS-R form compared to the GSH form. It is also the case for residues 13 and 17 (see Figure S5). Therefore, the presence of GS-R ligand modifies the structure of hGSTA1 for some residues, which corresponds to an induced fit model of protein–ligand binding. On the opposite, for the comparison between the APO and GSH forms, the population shift model of protein–ligand binding is more pronounced since the similarity between FESs and FEPs is larger than comparisons with GS-R forms (Figure 4A,B).

Finally, 10 FESs of CGAs were identified as moderately or largely modified due to ligand binding for the comparison between GS-R and APO forms of hGSTA1. In detail, CGAs $(\theta, \gamma)_i$ for $i = 10, 12, 13, 14, 15, 16, 17, 18, 53$, and 158 were identified. Among these 10 CGAs, six were already detected from GSH vs. GS-R comparison (Table S1). It means that these six CGAs are GS-R specific, i.e., they are similar in the APO and GSH forms compared to the GS-R form. Particularly, CGA $(\theta, \gamma)_{53}$, which involves residues Gln54 and Val55, is of special interest since both residues belong to the G site. In addition, CGA $(\theta, \gamma)_{158}$, which involves residue His159, is also of special interest since this residue belongs to the H site. In total, from the three different comparisons of ligand-binding forms of hGSTA1 shown in Figure 4A, 12 unique CGAs were identified to be influenced by the absence/presence of a ligand; some of them being detected in two different comparisons (Table S1). It corresponds to a total of 26 different residues with four residues which belong to the G site (Tyr9, Arg15, Gln54, and Val55) and four residues which belong to the H site (Phe10, Gly14, Ser18, and His159). All 2D FESs of the 12 CGAs identified to be involved in the ligand binding process in hGSTA1 are presented in Figure S5.

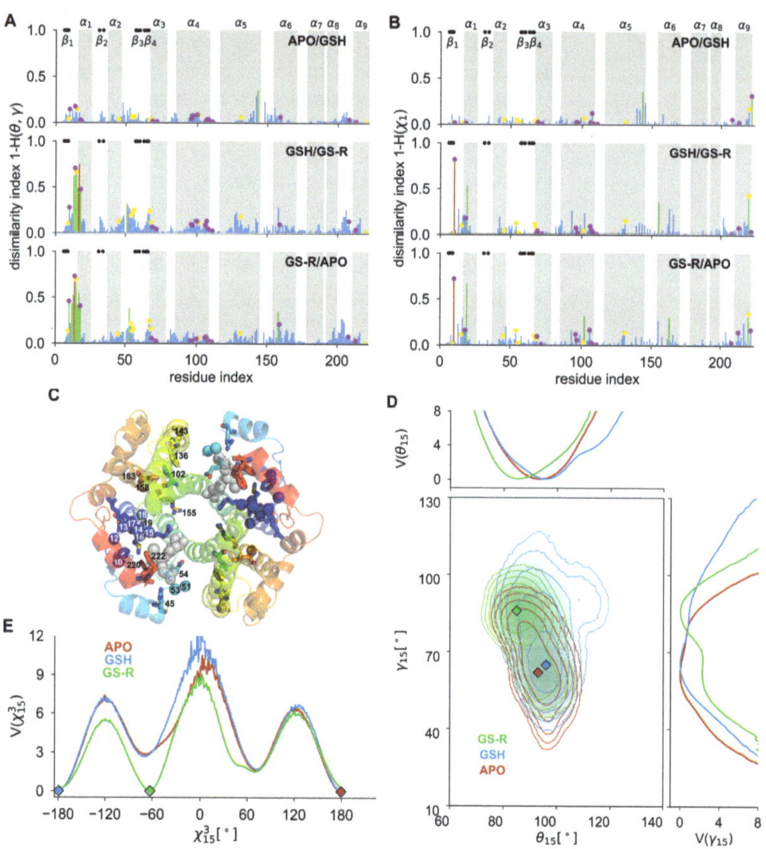

Figure 4. (**A**) Dissimilarity index $1 - H$ along the amino-acid sequence computed from FESs of CGAs (θ, γ). The color code is the following: low dissimilarity: blue, moderate dissimilarity: green, large dissimilarity: red (see Materials and Methods, Section 2.2). Yellow and purple circles correspond to residue that belongs to the G and H site, respectively. Gray rectangles indicate residues in α helix, and black circles at the top of each plot represent residues in β-sheets. (**B**) Dissimilarity index $1 - H$ along the amino-acid sequence computed from FEPs of SCAs χ^1. The color code is the same as in panel A. (**C**) Location of CGAs and SCAs detected from free-energy landscape analysis in the hGSTA1 structure. The color code is the following: N-term (blue) to C-term (red). (**D**) Effective 2D FESs $V(\theta, \gamma)_{15}$ in the APO, GSH, and GS-R forms of hGSTA1. Effective 1D FEPs of each internal coordinate θ_{15} and γ_{15} are also presented. (**E**) Effective FEPs $V(\chi^3)_{15}$ in the APO, GSH, and GS-R forms of hGSTA1. Diamonds indicate the global minimum in each form.

We applied the exact same procedure to compare the one-dimensional FEPs of SCAs χ^k. First, three amino acids were excluded from the *DATA* since they do not exhibit converged

FEPs of SCAs between monomers A and B of hGSTA1. It concerns residues Glu146 ($\chi^{1,2}$), His159($\chi^{1,2}$), and Arg221 ($\chi^{1,2,3,4,5}$). It represents only 1.5% of the total number of residues characterized by SCAs ($N = 192$). Then, from dissimilarity indices, $1 - H$, computed along the amino-acid sequence for the three different comparisons APO vs. GSH, GSH vs. GS-R, and GS-R vs. APO, as shown in Figure 4B for SCAs χ^1 and in Figure S3 for the other SCAs $\chi^{2,3,4,5}$, a total of 19 SCAs were identified to be influenced by the absence/presence of a ligand in the binding sites of hGSTA1 homodimer. It concerns 15 different residues, i.e., Phe10, Arg13, Arg15, Met16, Glu17, Thr19, Arg45, Gln54, Leu102, Phe136, His143, Arg155, Leu163, Phe220, and Phe222 (Table S1). Among these 15 residues, four are located in the G site (Arg15, Arg45, Gln54, and Phe220) and two in the H site (Phe10 and Phe222). In detail, the comparison between the APO and GSH forms of hGSTA1 highlights three SCAs, $\chi^{1,2}_{143}$ and χ^1_{222} (C-term). Particularly, residue His143 was also found to be crucial (the only one) for ligand binding from the analysis of the CGA (θ_i, γ_i) FESs. It means that residue His143 is an unexpected probe of ligand binding as both its local main chain conformation and its side chain are modified by the absence (APO) or presence (GSH) of the ligand, even though His143 is not directly located in the G or H binding site. In fact, this residue is at the surface of the hGSTA1 enzyme, at a distance around 30 Å from the binding sites.

Furthermore, 11 SCAs are found to show free-energy dissimilarities between their GSH and GS-R forms, i.e., $\chi^{1,2}_{10}$, χ^3_{15}, χ^3_{16}, χ^1_{19}, χ^3_{45}, χ^2_{54}, χ^2_{136}, $\chi^{1,3}_{155}$, and χ^1_{220}. The six residues mentioned above for the comparison of APO vs. GSH forms and which belong to the G and H sites were also identified above from the comparison of CGA FESs for APO vs. GSH. In particular, SCA χ^3_{15} is associated with Arg15 which is conserved in GST class Alpha and located at the interface between G and H sites. For this SCA, as shown in Figure 4E, the presence of GS-R ligand modifies the position of the global minimum of the FEP, from $-180°$ in both APO and GSH forms to $-60°$. This local conformation exists in APO and GSH forms but as a local minimum (population shift). Finally, 12 SCAs were detected for the comparison GS-R vs. APO. Among these 12 SCAs, seven were already identified from the comparison GSH vs. GS-R, i.e., residues Phe10, Arg15, Met16, Thr19, Phe136, and Phe220 are, therefore, GS-R specific. The same kind of behavior was already observed for CGAs of the main chain with six coordinates GS-R specific. Among these six coordinates, residue Arg15 was identified, which support its case as one of the most important residues involved in ligand binding of hGSTA1. The five supplementary SCAs identified are χ^3_{13}, χ^3_{17}, χ^1_{102}, and $\chi^{1,2}_{163}$, which correspond to residues Arg13, Glu17, Leu102, and Leu163. Particularly, Leu102 is located very close to the G site, Asp101 belonging to the G site and located in the opposite monomer of the dimer. All 1D FEPs of the 19 SCAs identified to be involved in the ligand binding process in hGSTA1 are presented in Figure S6.

3.3. Collective Motions of the Network of Residues Involved in Ligand Binding in hGSTA1 Revealed by PCA

As described in detail above, the free-energy landscape (FEL) comparison of CGAs (θ, γ) reveals that the absence/presence of ligands in hGSTA1 modifies at least 12 internal coordinates, located in both binding sites G and H or even further in the sequence (Figure 4C). In other words, the modification of the (unknown) free-energy landscape of hGSTA1 upon ligand binding can be represented here quantitatively by its projection over 12 dimensions. Obviously, these 12 coordinates are not independent and the local motions of the CGAs are coupled to each other. Hereafter, we establish the relation between these 12 local conformational changes by applying principal component analysis (see Supplementary Materials for details). From PCA for CGAs (θ, γ) in the three different forms of hGSTA1, we found that more than 30% of the total contribution to the MSF of the 12 CGAs is captured by the first two collective modes over the 72 existing ones (Figure S7). Then, we computed time series of the projections of vectors q_i along the eigenvectors of collective modes 1 and 2, which defined the collective coordinates PC_1 and PC_2. From the two-dimensional probability density functions of PC_1 and PC_2 time series computed from

MD trajectories, the effective 2D FEL $V_{1,2}$, shown in Figure 5, was computed for the three forms APO, GSH, and GS-R. The largest motions defined by the first and second PCs can be interpreted in terms of specific features of the two-dimensional FEL $V_{1,2}$. The 2D FEL of the APO form of hGSTA1 presents two minima defined from the contour lines of free-energy ($<2\,k_BT$), whereas both GSH and GS-R forms present three minima. In addition, for GSH and GS-R forms, the different minima are well separated by energy barriers from around 3 to 4 k_BT, whereas in the APO form, the two minima are separated by a barrier around 1 k_BT. These 2D FELs are, in fact, in agreement with the observation that FEPs of θ and γ angles (Figures 4D and S5) are more spread out and therefore contain more wells for GSH and GS-R forms in addition to the one(s) of the APO form.

From the 2D FEL shown in Figure 5A, the coupling between the 12 CGAs of the network identified as important for ligand binding in hGSTA1 was estimated by computing the influences of each coordinate into the largest amplitude collective motions, which are of particular interest for the biological function of proteins [34]. First of all, as shown in Figure 5B for each form, the dynamics of the 12 coordinates are highly asymmetrical between monomers A and B of hGSTA1. For collective mode 1 of APO, GSH, and GS-R forms, the contribution of monomer A is 66, 24, and 13%, respectively (with 34, 76, and 87% for monomer, B). For collective mode 2, the contribution of monomer A is even larger. Moreover, for the collective mode 1 of GSH form, the Pearson correlation of influences along the 12 CGAs is large, i.e., $\rho_{A|B} = 0.84$, whereas for the collective mode 2, the correlation is low, i.e., $\rho_{A|B} = 0.23$. However, PCA reveals two major dynamical couplings between residues. First, there is a coupling between CGAs 13–14–15 (including residues Gly14 of the H site and Arg15 of the G site) and CGAs 51, 53 (including residues Gln54 and Val55 of the G site) in the same monomer A or B. This coupling can be more or less pronounced between monomers and is modified by the presence of GS-R ligand, which modifies the location of CGA fluctuations in collective mode 1 (Figure 5B). Second, there is a coupling between the two binding sites of the two monomers, as shown particularly for CGAs 51, 53 in collective mode 1 for both APO and GSH form. This coupling is also present but is less pronounced in collective mode 2 of GS-R form. Therefore, dynamical coupling of intra-binding sites and inter-binding sites of hGSTA1 along the main chain of the enzyme was found and appears to be crucial for ligand binding in hGSTA1.

We applied the same procedure to the 19 SCAs detected from FEPs similarities. First, the two collective modes 1 and 2 contribute to 17, 18, and 36% for APO, GSH, and GS-R forms of hGSTA1, respectively (Figure S6). As shown in Figure 5C, 2D FELs $V_{1,2}$ from SCAs present much more minima than for CGAs. It comes from the fact that SCAs show unrestricted FEPs with values between -180 and $+180°$ compared to CGAs. Moreover, the shape of the FEL for the GS-R form of hGSTA1 is much more different compared to the APO and GSH forms. The GS-R form of hGSTA1 explores the free-energy landscape less and is more restricted. However, the free-energy barriers between minima are much larger than the APO and GSH forms (up to 6 k_BT), for which the multiple minima are more easily accessible to the system. This behavior is related to the fact that the binding of GS-R ligand, which is bigger than GSH ligand, involves more constraints on the side chains of the residues and, therefore, larger free-energy barriers to overcome. In addition, compared to CGAs, collective modes 1 and 2 are more spread out over the two monomers, particularly in the GSH form (Figure 5D). The analysis of the influences of each SCA in the collective motions of mode 1 and 2 reveals a strong coupling Leu163 and residue Glu17 and its neighbors in the G site in the APO form. In the GS-R form, the coupling is more pronounced with residue Arg155.

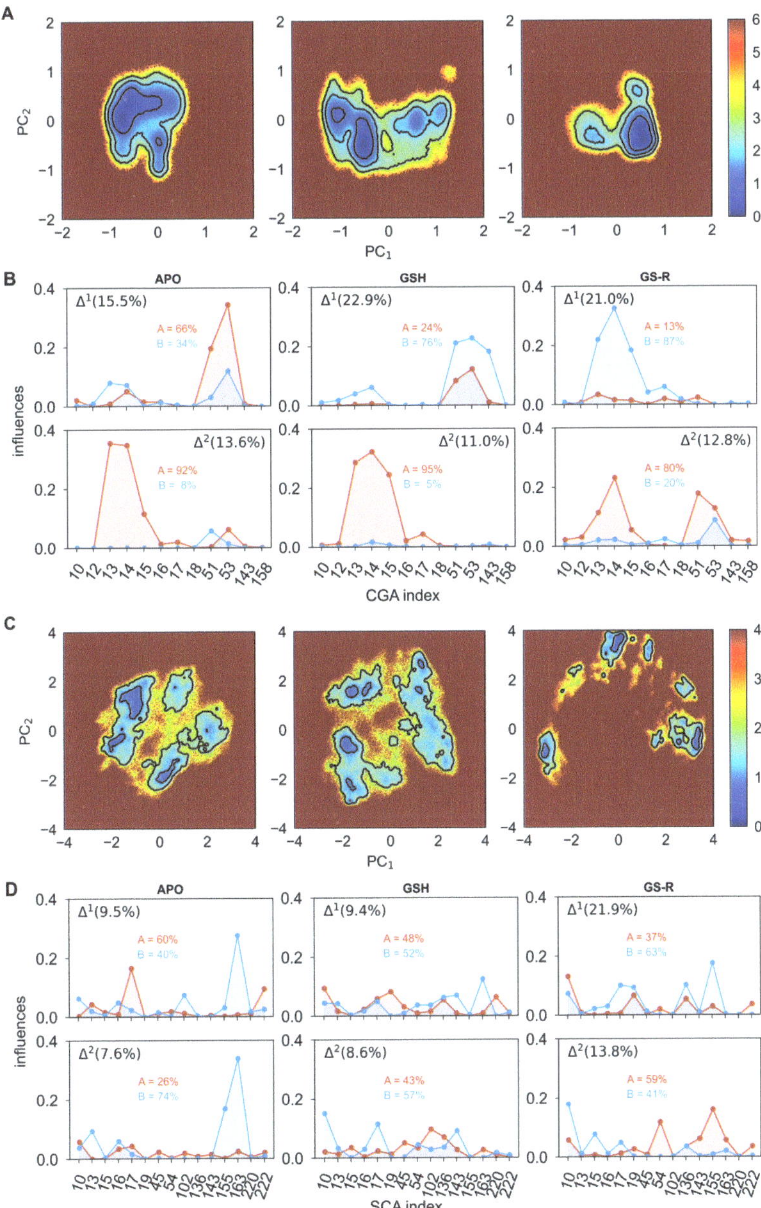

Figure 5. (**A**) 2D FEL computed from PCA applied on CGAs for hGSTA1 in its APO (**left** panel), GSH (**center** panel), and GS-R (**right** panel) forms. Contours (black lines) are drawn every k_BT. The color scale for the free-energy is in k_BT unit. (**B**) Influences Δ_i^k as a function of CGAs i for collective modes $k = 1, 2$ computed from PCA. Left panels concern the APO form, middle panels the GSH form, and right panels the GS-R form. Values in inset represent the percentage of contribution for the two monomers A and B of hGSTA1. (**C**) 2D FEL computed from PCA applied on SCAs for hGSTA1 in its APO (**left** panel), GSH (**center** panel), and GS-R (**right** panel) forms. (**D**) Influences Δ_i^k as a function of SCAs i for collective modes $k = 1, 2$ computed from PCA.

4. Discussion

From the three different comparisons of ligand-binding forms of human GSTA1 shown in Figure 1C, 12 CGAs and 19 SCAs were identified to be influenced by the absence/presence of ligands, most of them being gluthatione-S-conjugated-specific (Table S1). It corresponds to a total of 33 different amino acids (15% of the total sequence), eight of them being identified from both CGAs and SCAs analyses (Table 1). Among these 33 residues, 11 of them (32%) are directly located to the two binding sites of hGSTA1, with 6 which belong to the G site, i.e., Tyr9, Arg15, Arg45, Gln54, Val55, Phe220, and 5 which belong to the H site, i.e., Phe10, Gly14, Ser18, His159, and Phe222 (Figure S8). Furthermore, a collective motion involving residues Arg15, Gln54, and Val55 of the G site and residue Gly14 of the H site was identified using principal component analysis, revealing a strong coupling between these residues. Particularly, residue Arg15, which is found here to be involved in both local conformational changes of the main chain and of the side chains of hGSTA1, was demonstrated to have crucial catalytic activity both theoretically [35] and experimentally [36,37]. Finally, residues Phe220 (G site) and Phe222 (H site) belong to the C-terminal part of hGSTA1 (α_9-helix; see Figure S8) and it has been demonstrated that this secondary structure contributes remarkably to the catalytic and noncatalytic ligand-binding functions of GSTs [32,33].

Table 1. Summary of amino acids involved in the ligand binding to hGSTA1 and extracted from similarity of FESs of CGAs and FEP of SCAs. Amino acids, which were detected from both analyses, are underlined. Amino acids, which are highly conserved among GSTs, are denoted in bold. Secondary structure location of amino acids are indicated in parentheses.

Location	List of Amino Acids
G site	Tyr9 (β_1), <u>Arg15</u> (L_{β_1,α_1}), Arg45 (α_2), <u>Gln54</u>, **Val55** (L_{α_2,β_3}), Phe220 (α_9)
H site	<u>Phe10</u>, Gly14 (L_{β_1,α_1}), Ser18 (α_1), His159 (α_6), Phe222 (α_9)
Others	Asn11, Ala12, **Arg13** (L_{β_1,α_1}), <u>Met16</u>, <u>Glu17</u>, <u>Thr19</u>, **Arg20** (α_1), Leu50, Met51, Phe52, Gln53 ($\overline{L_{\alpha_2,\beta_3}}$), Leu102 ($\alpha_4$), Phe136, Ser142, <u>His143</u>, Gly144, Gln145 (α_5), Arg155, **Asp157**, Ile158, Leu160, Leu163 (α_6)

At large distance from the ligand-binding sites of hGSTA1, 12 residues (from Leu102 to Leu163; see Table 1) were identified as sensitive to the ligand disturbance. For example, His143 is involved in the main-chain and side-chain conformational changes of hGSTA1 and is located around 30 Å away from the G and H sites Figure S8). Moreover, GST sequence analysis indicates that some residues form an highly conserved local sequence GXXh(T/S)XXDh (h: hydrophobic), constituted by the α_6-helix and its preceding loop [38–40]. This motif belongs to a substructure named N-capping box and hydrophobic staple motif of GST and has been shown to be critical for the protein folding and stability [41]. In the present work, six residues were identified from the α_6-helix, particularly Arg155, Asp157 (highly conserved), and Ile158 (hydrophobic) which belongs to the corresponding motif GNKLSRADI. According to experimental structures, Ile158 can form hydrophobic interactions with the α_1-helix [41], which is an important structural element supporting the active binding site of GSTs.

Finally, it has been reported that the sequence alignment of all known GST structures (more than 100) shows that only 6–7 residues, i.e., less than 5% of the entire polypeptide chain, are strictly conserved [42]. For instance, Pro56 of the GST Alpha class, which belongs to the G site, is conserved through GST classes Mu, Pi, Sigma, Phi, Tau, Theta, Zeta, Omega, and Beta. In the present work, four amino acids, i.e., Arg13, Arg20, Val55, and Asp157, were detected and are highly conserved among GST classes. Particularly, Arg20 is strictly conserved through GST classes Alpha, Mu, Pi, Sigma, Tau, Zeta, and Omega; Arg13 is strictly conserved through GST classes Alpha, Mu, and Pi [43].

5. Conclusions

In the present work, we performed all-atom classical MD simulations in explicit solvent of human GSTA1 enzyme in the three different following forms of the protein: (i) the APO form, when no ligand is bound to hGSTA1, (ii) the GSH form, when the gluthatione ligand is bound to the G site of hGSTA1, and (iii) the GS-R form, when the gluthatione-S-conjugate is bound to the G and H sites of hGSTA1. From MD runs, we performed a free-energy landscape analysis of internal coordinates along the amino-acid sequence of the protein. The analysis was conducted using coarse-grained angles (θ, γ) of the main-chain (CGAs) and side-chain angles χ (SCAs), as local probe to track conformational changes. The comparison between each form of hGSTA1 reveals 12 CGAs and 19 SCAs influenced by the ligand binding into the G and H sites. It corresponds to a total of 33 residues of hGSTA1 which are involved in the protein–ligand binding process. Among the 33 residues identified, 11 of them belong to the two binding sites of hGSTA-1 identified experimentally from XRD structures, which confirms that the method developed and applied here is very robust. Finally, the dynamics of coarse-grained and side-chain angles were studied using principal component analysis. It shows an asymmetrical behavior between the two monomers of hGSTA1, particularly from CGAs. Moreover, the analysis of the first collective motions reveals a strong dynamical coupling between residues Arg15-Gln54-Val55 of the G site and Gly14 of the H site. These residues are coupled in the same binding site (intrasite) but also between the two binding sites of each monomer of hGSTA1 (intersites), which is an important result of the present work. Finally, residue Glu17 shows important coupling with residues Arg155 and Leu163, depending on the binding form of hGSTA1 (APO and GS-R, respectively). This free-energy landscape methodology, already applied successfully to chaperone proteins [28] and now to protein–ligand binding of human GSTA1 enzyme, is very powerful to identify network of residues and their dynamical coupling involved in the communication between domains or binding sites of biological systems. An important outcome of the present approach is its ability to decipher long-range effects, i.e., influence of residues far from the binding site on the ligand–protein association. This might guide engineered mutagenesis to tailor specific ligand–protein interactions.

Supplementary Materials: The following are available online at https://www.mdpi.com/article/10.3390/app12168196/s1, Section S1: details about MD protocol, structural stability of hGSTA1 dimer and monomers, ligand association/dissociation, and principal component analysis applied to internal angles coordinates. Figure S1 : Identification of residues which belong to the G and H site of hGSTA1 from PDB structures. Figure S2: RMSD of hGSTA1 structures in the APO, GSH, and GS-R forms computed from MD trajectories. Figure S3: Distance between ligands and binding site G of hGSTA1 computed from MD trajectories. Figure S4: Dissimilarity index along the amino-acid sequence for side-chain angles χ. Figure S5: 2D free-energy surfaces $V(\theta, \gamma)$ for all coarse-grained angles identified to be influenced by ligand binding in hGSTA1. Figure S6: 1D free-energy profiles $V(\chi)$ for all side-chain dihedral angles identified to be influenced by ligand binding in hGSTA1. Figure S7: Contribution of collective modes extracted from PCA to the total fluctuations observed in MD. Figure S8: Cartoon structure of hGSTA1, with the network of residues relevant for ligand binding highlighted. Table S1: Summary of CGAs and SCAs involved in the ligand binding to hGSTA1 and extracted from similarity of FESs and FEPs. References [44–57] are cited in the supplementary materials.

Author Contributions: Conceptualization, A.N. and P.S.; methodology, A.N.; validation, A.N. and N.P.; formal analysis, P.G., N.P., and A.N.; software P.D.; data curation, P.G., N.P., and A.N.; writing—original draft preparation, A.N.; writing—review and editing, P.G., P.D., F.N., and P.S.; supervision, P.S.; project administration, P.S.; funding acquisition, P.S. All authors have read and agreed to the published version of the manuscript.

Funding: This work is supported by the EIPHI Graduate School (contrat ANR-17-EUR-0002) and the Conseil Régional de Bourgogne Franche-Comté.

Institutional Review Board Statement: Not applicable.

Informed Consent Statement: Not applicable.

Data Availability Statement: The data presented in this study are available in the article and supplementary materials.

Acknowledgments: The simulations were performed using HPC resources from DSI-CCuB (Université de Bourgogne).

Conflicts of Interest: The authors declare no conflict of interest.

References

1. Du, X.; Li, Y.; Xia, Y.L.; Ai, S.M.; Liang, J.; Sang, P.; Ji, X.L.; Liu, S.Q. Insights into Protein–Ligand Interactions: Mechanisms, Models, and Methods. *Int. J. Mol. Sci.* **2016**, *17*, 144. [CrossRef] [PubMed]
2. Li, L.; Koh, C.C.; Reker, D.; Brown, J.B.; Wang, H.; Lee, N.K.; Liow, H.h.; Dai, H.; Fan, H.M.; Chen, L.; et al. Predicting protein-ligand interactions based on bow-pharmacological space and Bayesian additive regression trees. *Sci. Rep.* **2019**, *9*, 7703. [CrossRef] [PubMed]
3. Silva, J.L.; Vieira, T.C.R.G.; Gomes, M.P.B.; Bom, A.P.A.; Lima, L.M.T.R.; Freitas, M.S.; Ishimaru, D.; Cordeiro, Y.; Foguel, D. Ligand Binding and Hydration in Protein Misfolding: Insights from Studies of Prion and p53 Tumor Suppressor Proteins. *Acc. Chem. Res.* **2010**, *43*, 271–279. [CrossRef] [PubMed]
4. Payandeh, J.; Volgraf, M. Ligand binding at the protein–lipid interface: strategic considerations for drug design. *Nat. Rev. Drug Discov.* **2021**, *20*, 710–722. [CrossRef]
5. Chakraborti, S.; Hatti, K.; Srinivasan, N. 'All That Glitters Is Not Gold': High-Resolution Crystal Structures of Ligand-Protein Complexes Need Not Always Represent Confident Binding Poses. *Int. J. Mol. Sci.* **2021**, *22*, 6830. [CrossRef]
6. Mannervik, B. The isoenzymes of glutathione transferase. *Adv. Enzymol. Relat. Areas Mol. Biol.* **1985**, *57*, 357–417. [CrossRef]
7. Armstrong, R.N. Structure, Catalytic Mechanism, and Evolution of the Glutathione Transferases. *Chem. Res. Toxicol.* **1997**, *10*, 2–18. [CrossRef]
8. Hayes, J.D.; Flanagan, J.U.; Jowsey, I.R. Glutathione transferases. *Annu. Rev. Pharmacol. Toxicol.* **2005**, *45*, 51–88. [CrossRef]
9. Booth, J.; Boyland, E.; Sims, P. An enzyme from rat liver catalysing conjugations with glutathione. *Biochem. J.* **1961**, *79*, 516–524. [CrossRef]
10. Combes, B.; Stakelum, G.S. A liver enzyme that conjugates sulfobromophthalein sodium with glutathione. *J. Clin. Investig.* **1961**, *40*, 981–988. [CrossRef]
11. Axarli, I.; Rigden, D.; Labrou, N. Characterization of the ligandin site of maize glutathione S-transferase I. *Biochem. J.* **2004**, *382*, 885–893. [CrossRef] [PubMed]
12. Oakley, A. Glutathione transferases: A structural perspective. *Drug. Metab. Rev.* **2011**, *43*, 138–151. [CrossRef]
13. Mannervik, B.; Awasthi, Y.C.; Board, P.G.; Hayes, J.D.; Di Ilio, C.; Ketterer, B.; Listowsky, I.; Morgenstern, R.; Muramatsu, M.; Pearson, W.R. Nomenclature for human glutathione transferases. *Biochem. J.* **1992**, *282*, 305–306. [CrossRef] [PubMed]
14. Dirr, H.; Reinemer, P.; Huber, R. X-ray crystal structures of cytosolic glutathione S-transferases. *Eur. J. Biochem.* **1994**, *220*, 645–661. [CrossRef] [PubMed]
15. Lien, S.; Gustafsson, A.; Andersson, A.K.; Mannervik, B. Human Glutathione Transferase A1-1 Demonstrates Both Half-of-the-sites and All-of-the-sites Reactivity. *J. Biol. Chem.* **2001**, *276*, 35599–35605. [CrossRef]
16. Bocedi, A.; Fabrini, R.; Bello, M.L.; Caccuri, A.M.; Federici, G.; Mannervik, B.; Cornish-Bowden, A.; Ricci, G. Evolution of Negative Cooperativity in Glutathione Transferase Enabled Preservation of Enzyme Function. *J. Biol. Chem.* **2016**, *291*, 26739–26749. [CrossRef]
17. Fabrini, R.; De Luca, A.; Stella, L.; Mei, G.; Orioni, B.; Ciccone, S.; Federici, G.; Lo Bello, M.; Ricci, G. Monomer-Dimer Equilibrium in Glutathione Transferases: A Critical Re-Examination. *Biochemistry* **2009**, *48*, 10473–10482. [CrossRef]
18. Frova, C. Glutathione transferases in the genomics era: new insights and perspectives. *Biomol. Eng.* **2006**, *23*, 149–169. [CrossRef]
19. Board, P.G.; Menon, D. Glutathione transferases, regulators of cellular metabolism and physiology. *Biochim. Biophys. Acta* **2013**, *1830*, 3267–3288. [CrossRef]
20. Atkinson, H.J.; Babbitt, P.C. An Atlas of the Thioredoxin Fold Class Reveals the Complexity of Function-Enabling Adaptations. *PLoS Comput. Biol.* **2009**, *5*, e1000541. [CrossRef]
21. Deponte, M. Glutathione catalysis and the reaction mechanisms of glutathione-dependent enzymes. *Biochim. Biophys. Acta* **2013**, *1830*, 3217–3266. [CrossRef] [PubMed]
22. Mannervik, B.; Danielson, U.H. Glutathione transferases–Structure and catalytic activity. *CRC Crit. Rev. Biochem.* **1988**, *23*, 283–337. [CrossRef] [PubMed]
23. Cummins, I.; Dixon, D.P.; Freitag-Pohl, S.; Skipsey, M.; Edwards, R. Multiple roles for plant glutathione transferases in xenobiotic detoxification. *Drug. Metab. Rev.* **2011**, *43*, 266–280. [CrossRef] [PubMed]
24. Wilce, M.C.; Parker, M.W. Structure and function of glutathione S-transferases. *Biochim. Biophys. Acta* **1994**, *1205*, 1–18. [CrossRef]
25. Nicolaï, A.; Delarue, P.; Senet, P. Intrinsic Localized Modes in Proteins. *Sci. Rep.* **2015**, *5*, 18128. [CrossRef]
26. Grassein, P.; Delarue, P.; Nicolaï, A.; Neiers, F.; Scheraga, H.A.; Maisuradze, G.G.; Senet, P. Curvature and Torsion of Protein Main Chain as Local Order Parameters of Protein Unfolding. *J. Phys. Chem. B* **2020**, *124*, 4391–4398. [CrossRef]

27. Maisuradze, G.G.; Senet, P.; Czaplewski, C.; Liwo, A.; Scheraga, H.A. Investigation of protein folding by coarse-grained molecular dynamics with the UNRES force field. *J. Phys. Chem. A* **2010**, *114*, 4471–4485. [CrossRef]
28. Nicolaï, A.; Delarue, P.; Senet, P. Decipher the Mechanisms of Protein Conformational Changes Induced by Nucleotide Binding through Free-Energy Landscape Analysis: ATP Binding to Hsp70. *PLoS Comput. Biol.* **2013**, *9*, e1003379. [CrossRef]
29. Senet, P.; Maisuradze, G.G.; Foulie, C.; Delarue, P.; Scheraga, H.A. How main-chains of proteins explore the free-energy landscape in native states. *Proc. Natl. Acad. Sci. USA* **2008**, *105*, 19708–19713. [CrossRef]
30. Cote, Y.; Senet, P.; Delarue, P.; Maisuradze, G.G.; Scheraga, H.A. Anomalous diffusion and dynamical correlation between the side chains and the main chain of proteins in their native state. *Proc. Natl. Acad. Sci. USA* **2012**, *109*, 10346–10351. [CrossRef]
31. Guzzo, A.; Delarue, P.; Rojas, A.; Nicolaï, A.; Maisuradze, G.G.; Senet, P. Missense Mutations Modify the Conformational Ensemble of the alpha-Synuclein Monomer Which Exhibits a Two-Phase Characteristic. *Front. Mol. Biosci.* **2021**, *8*, 786123. [CrossRef] [PubMed]
32. Wu, B.; Dong, D. Human cytosolic glutathione transferases: structure, function, and drug discovery. *Trends Pharmacol. Sci.* **2012**, *33*, 656–668. [CrossRef] [PubMed]
33. Honaker, M.T.; Acchione, M.; Zhang, W.; Mannervik, B.; Atkins, W.M. Enzymatic detoxication, conformational selection, and the role of molten globule active sites. *J. Biol. Chem.* **2013**, *288*, 18599–18611. [CrossRef] [PubMed]
34. Berendsen, H.J.; Hayward, S. Collective protein dynamics in relation to function. *Curr. Opin. Struct. Biol.* **2000**, *10*, 165–169. [CrossRef]
35. Dourado, D.F.A.R.; Fernandes, P.A.; Mannervik, B.; Ramos, M.J. Glutathione transferase A1-1: Catalytic importance of arginine 15. *J. Phys. Chem. B* **2010**, *114*, 1690–1697. [CrossRef]
36. Björnestedt, R.; Stenberg, G.; Widersten, M.; Board, P.G.; Sinning, I.; Alwyn Jones, T.; Mannervik, B. Functional significance of arginine 15 in the active site of human class alpha glutathione transferase A1-1. *J. Mol. Biol.* **1995**, *247*, 765–773. [CrossRef]
37. Gildenhuys, S.; Dobreva, M.; Kinsley, N.; Sayed, Y.; Burke, J.; Pelly, S.; Gordon, G.P.; Sayed, M.; Sewell, T.; Dirr, H.W. Arginine 15 stabilizes an S(N)Ar reaction transition state and the binding of anionic ligands at the active site of human glutathione transferase A1-1. *Biophys. Chem.* **2010**, *146*, 118–125. [CrossRef]
38. Aceto, A.; Dragani, B.; Melino, S.; Allocati, N.; Masulli, M.; Di Ilio, C.; Petruzzelli, R. Identification of an N-capping box that affects the alpha 6-helix propensity in glutathione S-transferase superfamily proteins: a role for an invariant aspartic residue. *Biochem. J.* **1997**, *322 Pt 1*, 229–234. [CrossRef]
39. Dragani, B.; Stenberg, G.; Melino, S.; Petruzzelli, R.; Mannervik, B.; Aceto, A. The conserved N-capping box in the hydrophobic core of glutathione S-transferase P1-1 is essential for refolding. Identification of a buried and conserved hydrogen bond important for protein stability. *J. Biol. Chem.* **1997**, *272*, 25518–25523. [CrossRef]
40. Stenberg, G.; Dragani, B.; Cocco, R.; Mannervik, B.; Aceto, A. A Conserved "Hydrophobic Staple Motif" Plays a Crucial Role in the Refolding of Human Glutathione Transferase P1-1. *J. Biol. Chem.* **2000**, *275*, 10421–10428. [CrossRef]
41. Cocco, R.; Stenberg, G.; Dragani, B.; Principe, D.R.; Paludi, D.; Mannervik, B.; Aceto, A. The Folding and Stability of Human Alpha Class Glutathione Transferase A1-1 Depend on Distinct Roles of a Conserved N-capping Box and Hydrophobic Staple Motif. *J. Biol. Chem.* **2001**, *276*, 32177–32183. [CrossRef] [PubMed]
42. Atkinson, H.J.; Babbitt, P.C. Glutathione Transferases Are Structural and Functional Outliers in the Thioredoxin Fold. *Biochemistry* **2009**, *48*, 11108–11116. [CrossRef] [PubMed]
43. Stenberg, G.; Board, P.G.; Carlberg, I.; Mannervik, B. Effects of directed mutagenesis on conserved arginine residues in a human Class Alpha glutathione transferase. *Biochem. J.* **1991**, *274 Pt 2*, 549–555. [CrossRef]
44. Grahn, E.; Novotny, M.; Jakobsson, E.; Gustafsson, A.; Grehn, L.; Olin, B.; Madsen, D.; Wahlberg, M.; Mannervik, B.; Kleywegt, G.J. New crystal structures of human glutathione transferase A1-1 shed light on glutathione binding and the conformation of the C-terminal helix. *Acta Cryst. D* **2006**, *62*, 197–207. [CrossRef]
45. Abraham, M.J.; van der Spoel, D.; Lindahl, E.; Hess, B.; The GROMACS Development Team. *GROMACS User Manual Version 5.1.5*; GROMACS: Groningen, The Netherlands, 2017.
46. Best, R.B.; Hummer, G. Optimized Molecular Dynamics Force Fields Applied to the Helix-Coil Transition of Polypeptides. *J. Phys. Chem. B* **2009**, *113*, 9004–9015. [CrossRef] [PubMed]
47. Lindorff-Larsen, K.; Piana, S.; Palmo, K.; Maragakis, P.; Klepeis, J.L.; Dror, R.O.; Shaw, D.E. Improved side-chain torsion potentials for the Amber ff99SB protein force field. *Proteins* **2010**, *78*, 1950–1958. [CrossRef]
48. Best, R.B.; de Sancho, D.; Mittal, J. Residue-Specific a-Helix Propensities from Molecular Simulation. *Biophys. J.* **2012**, *102*, 1462–1467. [CrossRef]
49. Jorgensen, W.L.; Chandrasekhar, J.; Madura, J.D.; Impey, R.W.; Klein, M.L. Comparison of simple potential functions for simulating liquid water. *J. Chem. Phys.* **1983**, *79*, 926–935. [CrossRef]
50. Wang, J.; Wang, W.; Kollman, P.A.; Case, D.A. Automatic atom type and bond type perception in molecular mechanical calculations. *J. Mol. Graph. Model* **2006**, *25*, 247–260. [CrossRef]
51. Wang, J.; Wolf, R.M.; Caldwell, J.W.; Kollman, P.A.; Case, D.A. Development and testing of a general amber force field. *J. Comput. Chem.* **2004**, *25*, 1157–1174. [CrossRef]
52. Sousa da Silva, A.W.; Vranken, W.F. ACPYPE—AnteChamber PYthon Parser interfacE. *BMC Res. Notes* **2012**, *5*, 367. [CrossRef] [PubMed]

53. Hess, B.; Bekker, H.; Berendsen, H.J.C.; Fraaije, J.G.E.M. LINCS: A linear constraint solver for molecular simulations. *J. Comput. Chem.* **1997**, *18*, 1463–1472. [CrossRef]
54. Bussi, G.; Donadio, D.; Parrinello, M. Canonical sampling through velocity rescaling. *J. Chem. Phys.* **2007**, *126*, 014101. [CrossRef]
55. Parrinello, M.; Rahman, A. Polymorphic transitions in single crystals: A new molecular dynamics method. *J. Appl. Phys.* **1981**, *52*, 7182–7190. [CrossRef]
56. Darden, T.; York, D.M.; Pedersen, N.L. Particle mesh Ewald: An N.log(N) method for Ewald sums in large systems. *J. Chem. Phys.* **1993**, *98*, 10089. [CrossRef]
57. Altis, A.; Nguyen, P.H.; Hegger, R.; Stock, G. Dihedral angle principal component analysis of molecular dynamics simulations. *J. Chem. Phys.* **2007**, *126*, 244111. [CrossRef]

Article

Information Transfer in Active States of Human β$_2$-Adrenergic Receptor via Inter-Rotameric Motions of Loop Regions

Nuray Sogunmez and Ebru Demet Akten *

Department of Molecular Biology and Genetics, Faculty of Engineering and Natural Sciences, Kadir Has University, Istanbul 34083, Turkey
* Correspondence: demet.akten@khas.edu.tr; Tel.: +90-212-533-65-32 (ext. 1350)

Featured Application: Loop regions in β$_2$AR are critical hot spot regions, likely in other GPCRs, and can be used as potential allosteric drug targets.

Abstract: Two independent 1.5 μs long MD simulations were conducted for the fully atomistic model of the human beta2-adrenergic receptor (β$_2$AR) in a complex with a G protein to investigate the signal transmission in a fully active state via mutual information and transfer entropy based on α-carbon displacements and rotameric states of backbone and side-chain torsion angles. Significant correlations between fluctuations in α-Carbon displacements were mostly detected between transmembrane (TM) helices, especially TM5 and TM6 located at each end of ICL3 and TM7. Signal transmission across β$_2$-AR was quantified by shared mutual information; a high amount of correspondence was distinguished in almost all loop regions when rotameric states were employed. Moreover, polar residues, especially *Arg*, made the most contribution to signal transmission via correlated side-chain rotameric fluctuations as they were more frequently observed in loop regions than hydrophobic residues. Furthermore, transfer entropy identified all loop regions as major entropy donor sites, which drove future rotameric states of torsion angles of residues in transmembrane helices. Polar residues appeared as donor sites from which entropy flowed towards hydrophobic residues. Overall, loops in β$_2$AR were recognized as potential allosteric hot spot regions, which play an essential role in signal transmission and should likely be used as potential drug targets.

Keywords: transfer entropy; rotameric state; loop region; allosteric network; mutual information

Citation: Sogunmez, N.; Akten, E.D. Information Transfer in Active States of Human β$_2$-Adrenergic Receptor via Inter-Rotameric Motions of Loop Regions. *Appl. Sci.* **2022**, *12*, 8530. https://doi.org/10.3390/app12178530

Academic Editor: Rosanna Di Paola

Received: 29 June 2022
Accepted: 13 August 2022
Published: 26 August 2022

Publisher's Note: MDPI stays neutral with regard to jurisdictional claims in published maps and institutional affiliations.

1. Introduction

Allostery is an essential property of all proteins that is accepted to be intrinsic, irrespective of their functional requirements [1,2]. In fact, all proteins are dynamic entities, which sample distinct conformational states, and allostery is manifested as the shift in that conformational ensemble when one site of a protein is triggered by either a bound ligand or a mutation. In some proteins, the catalytic region at a distant site experiences a change in its functional capacity; thus, allostery becomes a critical part of the protein's functional regulation [2–4]. In addition to conformational changes, allostery can also manifest itself as a change in the global dynamics of the protein. Based on Cooper and Dryden's model proposed almost 30 years ago, allostery arises from the changes in frequencies and amplitudes of thermal fluctuations even in the absence of any conformational change in the backbone [5]. In this entropic model of allostery, there is no redistribution in the preexisting conformational substates. However, there is a change in the depth of the corresponding local minima in which a coordinated fluctuation of residues transmits the change from one site to another distant site [6–8].

Another important aspect of allostery is the pathway along which the residues fluctuating in the correspondence are distributed. These so-called "hot spot" residues are essential for site-to-site communication and are valuable for computer-aided drug design studies as they often provide high specificity/selectivity in comparison to orthosteric binding site residues, which are mostly conserved among species [9–14]. Over the years, several graph-based algorithms have been developed to estimate this functional allosteric circuit with its constitutive residues [15–17]. Here, we used correlated fluctuations between residues to establish an allosteric communication network, which is described in terms of entropy/information transfer from one site to another. Transfer entropy was previously introduced by Schreiber in 2000 as an information-theoretic measure to quantify the exchange of information between two systems [18] and was later used in several MD studies as an analysis tool to understand the effects of various structural changes [19–23].

In our current study, the transfer information of coupled fluctuations was not only based on translational Cα displacements, as often considered in several studies, but also on rotational displacements of backbone and side-chain torsion angles in each residue. The thermodynamic importance of side-chain variability was previously emphasized in calmodulin-ligand binding studies [24,25]. Furthermore, NMR mutational studies demonstrated the contribution of side-chain fluctuations to long-range communication networks [26]. Previously, using Monte Carlo sampling, DuBay and his coworkers demonstrated that allosteric communication in proteins can be transmitted by correlated side-chain fluctuations only [27]. However, they assumed a fixed backbone rotation and quantified the correspondence using a mutual information metric only. Here, our study will be the first to consider both backbone and side-chain rotatable bonds altogether to identify the correlated fluctuations in the rotameric states of these torsion angles. In addition, we will use another information-theoretic measure, the so-called *"transfer entropy"* to determine the dynamics of information transport, i.e., the direction of the exchange of information from one site to another distant site in the receptor at a future time.

The system under study is a human beta2-adrenergic receptor (β_2AR) in complex with G protein representing the active state. It was subjected to two separate 1.5 μs long MD simulations, which amounted to a 3 μs long trajectory. Dynamic cross-correlation analysis based solely on α-Carbon displacements was followed by mutual information and transfer entropy calculations based on fluctuations in both α-Carbon displacements and rotameric state of backbone and side-chain rotatable bonds. A significant amount of correspondence was observed for fluctuations in rotameric states for residues in loop regions. This overlooked information carried via fluctuations within the rotameric well was emphasized for the first time in this study as an important component of allosteric regulation. Furthermore, the information transfer was directed from polar residues located in loop regions towards hydrophobic residues found in the transmembrane regions of the receptor, i.e., fluctuations in rotameric states of polar loop residues dictated the future fluctuations of rotameric states of hydrophobic transmembrane residues. This driver–follower relation between the loop and transmembrane regions of the receptor via polar/hydrophobic residue pairs elucidated for the first time an important allosteric communication network that can be used for allosteric drug design studies.

2. Materials and Methods

System Preparation. The active state of human β_2AR in a complex with a Gs complex and bound to agonist BI-167107 with a PDB id of 3SN6 [28] was used as an initial state conformation for MD simulations. Prior to the runs, T4 Lysozyme, nanobody, and the agonist were removed, the missing extracellular (*Ala176-His178*) and intracellular (*Phe239-Phe265*) residues were completed via a MODELLER homology modeling tool [29], and the mutations T96M, T98M and E187N, which were used as linkers in crystal structure formation were reverted to their original state via the *mutate* plugin of the VMD visualization tool [30]. The system was then embedded into a palmitoyloleoyl-phosphatidylcholine (POPC) lipid bilayer using VMD's *membrane* plugin tool [30], solvated with TIP3P water molecules, and later ionized with 160 Na^{+2} and 154 Cl^{-1} counter ions for neutralization, which is necessary for the Particle-Mesh Ewald summation method. The system with the dimensions of 125 × 125 × 165 Å was prepared with a total of 228,299 atoms, of which 54,707 were water molecules. The CHARMM36 forcefield was used to describe the interaction potential of protein and lipids [31]. Periodic boundary conditions were applied in an isothermal, isobaric NPT ensemble with a constant temperature of 310 K and a constant pressure of 1 bar. Temperature and pressure were controlled by the Langevin thermostat and Langevin piston barostat, respectively [32]. The equations of motion were integrated with a 2 fs time step, and the SHAKE algorithm was used to constrain covalent bonds involving non-water hydrogen bonds with a non-bonded cutoff value of 12 Å.

Two independent 1.5 μs long MD runs were performed via the NAMD v2.13 software tool. Each run was initiated with three steps of initial energy minimizations under flexible cell conditions, including (i) the melting of lipid tails when the rest of the atoms were fixed, (ii) minimization and equilibration when protein was constrained but lipid, water, and ion atoms were released, and (iii) minimization and equilibration with the release of all atoms, which was then followed by equilibrium and production runs under constant area according to the membrane proteins' simulation protocol of NAMD [33]. The lipid bilayer in the system was continuously monitored in the minimization and equilibration steps until reaching 63.69 Å^2 area per lipid ratio, which was in the range of the experimentally reported value of 64.3 ± 1.3 Å^2 [34,35].

Dynamic Cross-Correlation. Correlations between atomic fluctuations from average positions of two residues i and j were calculated using the following equation:

$$C_{i,j} = C(\Delta R_i, \Delta R_j) = \frac{\langle \Delta R_i(t) \cdot \Delta R_j(t) \rangle}{\sqrt{\langle (\Delta R_i)^2 \rangle \langle (\Delta R_j)^2 \rangle}} \tag{1}$$

The time average of the dot product of $\Delta R_i(t)$ and $\Delta R_j(t)$ was taken and normalized. $\Delta R_i(t)$ and $\Delta R_j(t)$ represent the atomic fluctuations of α-Carbons only. If $C_{ij} = 1$, then the fluctuations of atoms i and j are perfectly correlated (fluctuates in the same direction), if $C_{ij} = -1$, then the fluctuations of atoms i and j are perfectly anticorrelated (fluctuates in opposite directions), and if $C_{ij} = 0$, then the atoms i and j fluctuate independently.

Contact map generation. The cutoff distance (R_c) for heavy atoms (C, N, O, S) was taken as 6 Å, below which the atoms were considered to be in contact. The incorporation of all heavy atoms provides a more accurate representation of the contact profile than that of α-Carbons only. The formula used for contact map calculation was defined as:

$$M_{i,j} = \begin{cases} 1, \ if \ \delta_{i,j} \leq R_c \\ 0, \ Otherwise \end{cases} \tag{2}$$

Contact percentages over the MD trajectory were calculated with $\sum_{n=1}^{nconf} M_{i,j}/nconf$ and the threshold was set to 75% of the whole trajectory to recognize stable contacts.

Mutual Information (MI). Mutual information based on α-Carbon positional fluctuations between residue pairs i and j was calculated using the following expression:

$$MI(i,j) = \sum_k p\big(\Delta R_i(t_k), \Delta R_j(t_l)\big) log_2 \frac{p\big(\Delta R_i(t_k), \Delta R_j(t_l)\big)}{p\big(\Delta R_i(t_k)\big) \cdot p\big(\Delta R_j(t_l)\big)} \tag{3}$$

where $p\big(\Delta R_i(t_k), \Delta R_j(t_l)\big)$ represents the joint probability of observing the fluctuation of residue i in state k and that of residue j in state l. Mutual information is a non-negative and symmetric quantity, and zero if the fluctuations of residue i are independent of the fluctuations of residue j. To calculate the probability of occurrence, $p\big(\Delta R_i(t_k), \Delta R_j(t_l)\big)$, the number of states k and l, also described as the number of bins, N_{bins}, were determined for each residue separately using Shannon's entropy criterion. The number of bins (or states) for each residue was determined as the value for which Shannon's entropy reaches its maximum. The convergence criterion was expressed as:

$$\frac{|H(N_{bins}+1)_i - H(N_{bins})_i|}{H(N_{bins})_i} < 0.02 \tag{4}$$

where $H(N_{bins})_i$ is the Shannon entropy for residue i with N_{bins}. Similarly, mutual information based on fluctuations in backbone and side torsion angles were expressed as:

$$MI_{i,j} = \sum_{\Theta_i}\sum_{\Theta_j} p(\Theta_i,\ \Theta_j)\ log_2\left(\frac{p(\Theta_i,\ \Theta_j)}{p(\Theta_i),\ p(\Theta_j)}\right) \tag{5}$$

where $p(\Theta_i,\ \Theta_j)$ denotes the joint probability of observing the joint state $(\Theta_i,\ \Theta_j)$ of residues i and j. Here, Θ_i and Θ_j represent the rotameric states of backbone φ, ψ and side-chain dihedrals χ_i, $i = 1,2,3,4$ in residues i and j, respectively. Based on the distribution of rotameric states, the number of discrete rotameric states (or bins) for backbone dihedrals was set to 3, whereas for side-chain dihedrals, the number of states varied between 0 and 6 according to the rotamer library [36].

Transfer Entropy. Transfer entropy is defined as the reduction in uncertainty in future states of residue j at $t + \tau$ by knowing the states of residue i at time t. Based on Shreiber's work [18], it was defined by Erman et al. [22,23] as;

$$TE_{i\rightarrow j}(\tau) = H\big(\Delta R_j(t)\big|\Delta R_j(t-\tau)\big) - H\big(\Delta R_j(t)\big|\Delta R_j(t-\tau),\Delta R_i(t-\tau)\big) \tag{6}$$

where $H\big(\Delta R_j(t)\big|\Delta R_j(t-\tau)\big)$ is the conditional entropy of residue j at time t given the values of ΔR_j at time $t - \tau$. The second term $H\big(\Delta R_j(t)\big|\Delta R_j(t-\tau),\Delta R_i(t-\tau)\big)$ is the conditional entropy of residue j at time t given the values of ΔR_i and ΔR_j at time $t - \tau$. When entropies are expressed as a function of the probability of occurrences of positional fluctuations ΔR, $TE_{i\rightarrow j}(\tau)$ becomes,

$$
\begin{aligned}
TE_{i\rightarrow j} = \ &-\langle log_2\ p(\Delta R_j(t),\Delta R_j(t-\tau))\rangle + \langle log_2\ p(\Delta R_j(t-\tau))\rangle \\
&+\langle log_2\ p(\Delta R_j(t),\ \Delta R_j(t-\tau),\ \Delta R_i(t-\tau))\rangle - \langle log_2\ p(\Delta R_j(t-\tau),\ \Delta R_i(t-\tau))\rangle
\end{aligned} \tag{7}
$$

A similar expression for transfer entropy was used for rotameric states, where ΔR was replaced by Θ, which includes the information of the rotameric state of all rotatable sp^3-sp^3 bonds in each residue $(\varphi,\ \psi,\chi_i,\ i = 1,2,3,4)$. However, for a residue pair such as *Lys-Arg*, where both residues include four side-chain rotatable bonds, each having 3 alternative rotameric states, the joint probability $p(\Theta_j(t),\ \Theta_j(t-\tau),\ \Theta_i(t-\tau))$ is comprised of 3^{18} $(= 3^6 \times 3^6 \times 3^6)$ different rotameric states. For a protein system with 312 residues, the calculation of the transfer entropy becomes computationally intractable as it exceeds the maximum size an array can hold. Thus, only the first side-chain rotameric state was considered together with two backbone torsion angles, i.e., $(\varphi,\ \psi,\chi_1)$, which yielded 3^9 $(= 3^3 \times 3^3 \times 3^3)$ different states per residue. Finally, the net transfer entropy was determined by taking the difference between TE from i to j and that from j to i as,

$$NetTE_{i\rightarrow j} = TE_{i\rightarrow j} - TE_{j\rightarrow i} \tag{8}$$

The source codes for both mutual information and transfer entropy calculations were written by the authors using C programming language and can be provided upon request.

3. Results and Discussion

The active state of the receptor was well preserved throughout the simulation. As all members of the G protein-coupled receptor (GPCR) superfamily, human β_2AR shares the 7TM structural motif, which consists of seven transmembrane-spanning alpha helices connected by loop regions at the intra- and extracellular sides of the membrane (See Figure 1a). Among other loop regions, the intracellular loop 3 (ICL3) plays a critical role in the recognition of G proteins [28]. In addition, there exists an allosteric coupling between ICL3 and the extracellular regions of the receptor, which incorporate the orthosteric binding site [37–39]. Conformational changes observed at the intracellular part affect the extracellular part, specifically the binding site, which holds key residues such as *Asp*113 on transmembrane helix 3 (TM3), *Ser*203, *Ser*204, *Ser*207 on TM5, *Phe*289, *Asn*293 on TM6, and *Asn*312 on TM7.

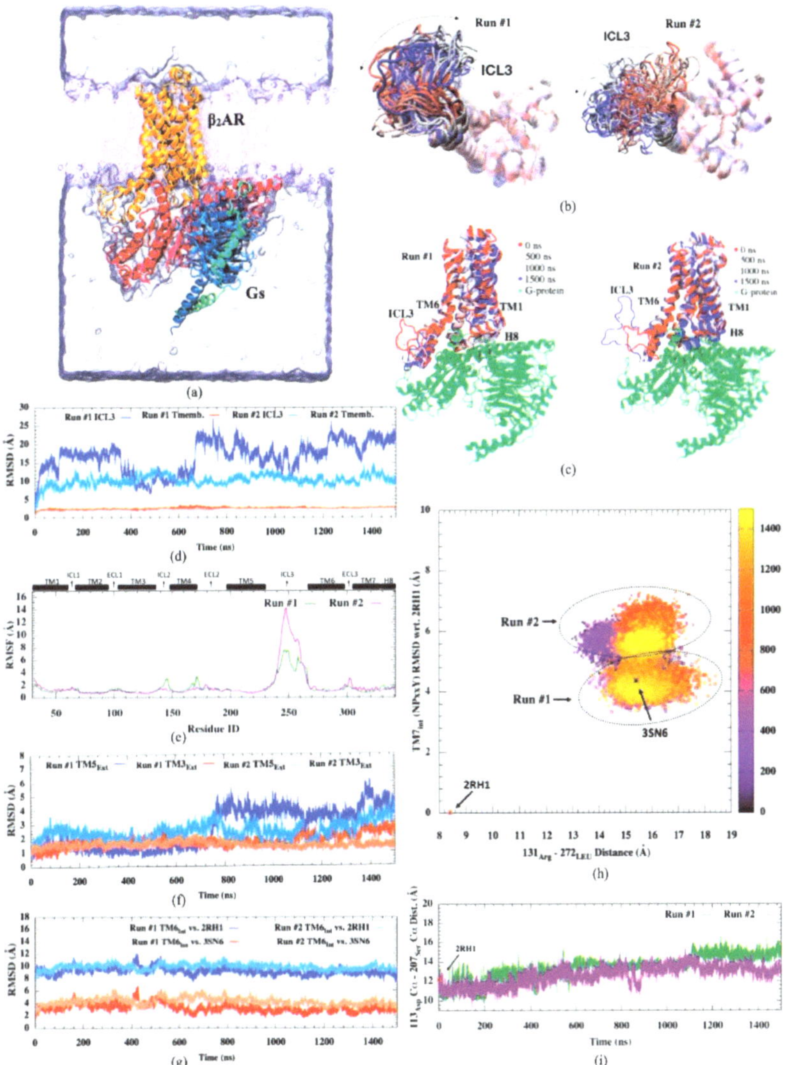

Figure 1. (**a**) Representation of human β_2AR in a complex with Gs complex in a membrane environment (created via VMD visualization tool [30]), (**b**) Intracellular views of 20 snapshots of ICL3 (**c**) side views of β_2AR-Gs complex for 4 different snapshots colored from red (initial) to white (intermediate), to blue (final) during simulation (**d**) RMSD profiles of ICL3 and transmembrane regions, (**e**) RMSF profiles (**f**) RMSD profile of extracellular parts of TM3 and TM5 helices, (**g**) RMSD profiles of intracellular parts of TM6, (**h**) RMSD of NPxxY motif of TM7 with respect to the inactive state (PDB id:2RH1) versus ionic lock distance ($Arg131C_\alpha$-$Leu272C_\alpha$), (**i**) distance profile for $Asp113C_\alpha$-$Ser207C_\alpha$ residue pair.

The most important conformational changes observed in two independent runs were summarized in Figure 1. The active state of the receptor was well characterized by an approximately 11 Å outward movement in the cytoplasmic end of TM6, and consequently, the adjoining ICL3 was pushed aside towards the lipid molecules (See Figure 1a–c). This unique conformation of the active state was only preserved in the presence of a G protein, which displaces TM6 and ICL3 outward for easy access to the receptor's binding cavity. Aligned snapshots of the receptor indicated high mobility in ICL3 in both runs, slightly enhanced in the second run. RMSD profiles of transmembrane helices 3, 5, 6, and 7 (TM3, TM5, TM6, and TM7) indicated that their initial states were well preserved throughout the simulation within the boundaries characterizing the active state of the receptor (Figure 1d–h). On the other hand, the distance between α-Carbons of two key residues at the orthosteric binding site, *Asp*113, and *Ser*207, displayed a slight increase from a range of 10–12 Å up to 13–16 Å in both runs, especially more noticeable in Run #1 (See Figure 1i). Our simulations were conducted with no ligand attached at the orthosteric binding site. Thus, the increase in the distance between these two key residues clearly indicated the tendency of the cavity to expand a bit in the absence of any agonist attached, irrespective of the fact that the active state was well preserved throughout the receptor, especially at the G protein binding site.

Previously, we attempted to simulate the active state in its free form, i.e., its G protein partner removed (PDB id: 3SN6), and observed that the initially opened and swept away ICL3 region and its adjoining transmembrane helix 6 (H6) swiftly changed position towards the core of the receptor at the very early stages of the simulation (in the first 50 ns), closed itself towards the core of the receptor, and blocked the G protein binding cavity. Moreover, in a simulation study conducted by Ozgur et al. [37], bond restraints were employed at the orthosteric binding site to preserve certain key distances between TM3 and TM5 within the experimentally reported range that represented the active state. No G protein was attached, yet ICL3 preserved its initial open conformation as if there was a G protein nearby, although the TM6's upward tilt characterizing the active state was not observed. Clearly, the conformational state of the orthosteric binding site allosterically affects the conformational state of the distant G protein binding site. It might facilitate the opening of the cavity for the initial binding of G protein, yet the fully active state can only be achieved and preserved when there is a G protein nearby interacting with the receptor. In other words, the major conformational shift for the characteristic tilt in TM6 requires an energy boost that a G protein can only provide and thus cannot be achieved in the course of an MD simulation, which is confined to low energy conformational states. Moreover, the absence of an agonist, which is accompanied by only a slight expansion at the unoccupied orthosteric binding site, does not destabilize the active state, which is already securely preserved by a G protein.

Cross-correlations between α-Carbon displacements disclosed TM6 as the dominating site fluctuating in concert with the rest. It is important to highlight distant regions that display positional fluctuations that are correlated with each other, as they might indicate the presence of some potential sites in communication along the allosteric pathway. Thus, the contact map was overlaid with the residue-pair cross-correlation map, as depicted in Figure 2. The contact map was generated by using the heavy atoms with a threshold distance of 6 Å. Distant and correlated regions were mainly detected in the second half of the receptor composed of TM5, TM6, ICL3, TM7, and its small extension H8, especially TM6, which incorporated most critical sites for binding intracellular G proteins and small extracellular molecules. Especially, *Phe*289, *Asn*293 on TM6, and *Asn*312 on TM7, which are known to be key residues interacting with the ligand at the orthosteric site, fluctuated in the same direction, with ICL3 having correlation values as high as 0.8. Another set of critical residues at the orthosteric binding sites *Ser*203, *Ser*204, and *Ser*207 located at the extracellular part of TM5 negatively correlated with the distant helical segment TM7 with a C_{ij} value of around -0.5. Specifically, it is interesting to observe both *Ser*203 and *Ser*207 on one side of the binding cavity fluctuating in opposite directions with *Asn*312 on the opposite side.

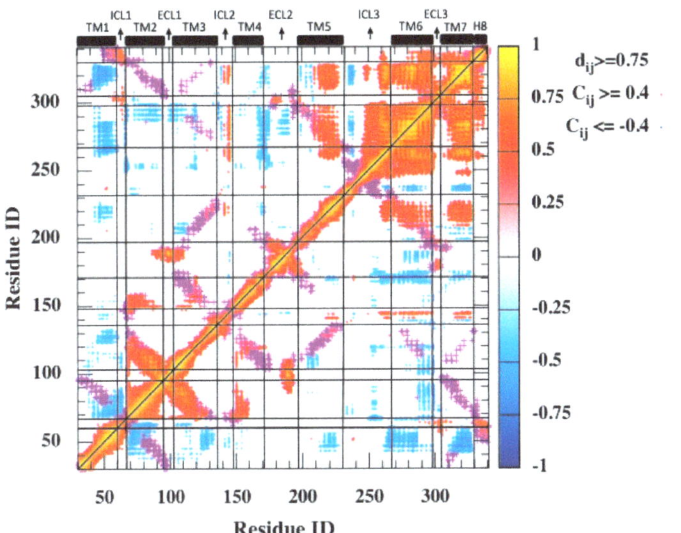

Figure 2. Residue-pair cross-correlation map for 3 μs MD trajectory. Only $C_{ij} > +0.4$ and $C_{ij} < -0.4$ are represented. Magenta dots represent the contact map (threshold distance of 6 Å).

Moreover, ICL2, the second-most important intracellular loop after ICL3, distantly fluctuated in concert with the second half of the receptor (TM5–H8). Finally, the first transmembrane helix TM1, which incorporated the free amino-terminal tail, mostly fluctuated in the opposite direction from the rest of the receptor, especially the distant helices TM6, TM7, and to some extent with TM3 and TM5, which all incorporated critical key residues at the orthosteric binding site.

Rotamer-based mutual information is mostly observed between loop regions and shared among polar residues. Cross-correlation is a metric that ignores the correlated motions in orthogonal directions. Therefore, even perfectly correlated motions important for allosteric signaling may be overlooked if the positional fluctuations are perpendicular to one another. On the other hand, mutual information, a metric in information theory, determines the correspondence between fluctuations of residue pairs, irrespective of their directions. MI was first calculated for the positional fluctuations of backbone Cα atoms, and as anticipated, the highest MI values (max. 4.01) were observed for residues close in space (see the diagonal line in Figure 3a). In addition, it is important to recognize high MI values observed between spatially distant residues as they would likely indicate the existence of an allosteric communication network, which is usually characterized by distant regions with a high degree of correspondence. As depicted in Figure 3a, mutual information was plotted together with a contact map to unravel the long-distance coupled motions of the allosteric network (see magenta dots). However, the only significant correspondence in distant Cα fluctuations was detected between a few residues located in ICL3's midpoint and the distant extracellular parts of TM6, extracellular loop 3 (ECL3), and TM7, which incorporate critical orthosteric binding site residues such as *Phe*289, *Asn*293, and *Asn*312. It is obvious that mutual information between backbone atomic fluctuation was mostly shared by neighboring residues either close in sequence or space. For all MI maps in Figure 3, red dots represent MI values greater than 0.5, and green dots represent MI between 0.25 and 0.5. Any MI less than 0.25 was not displayed. For clarity, the contact map was illustrated in Figure 3a only.

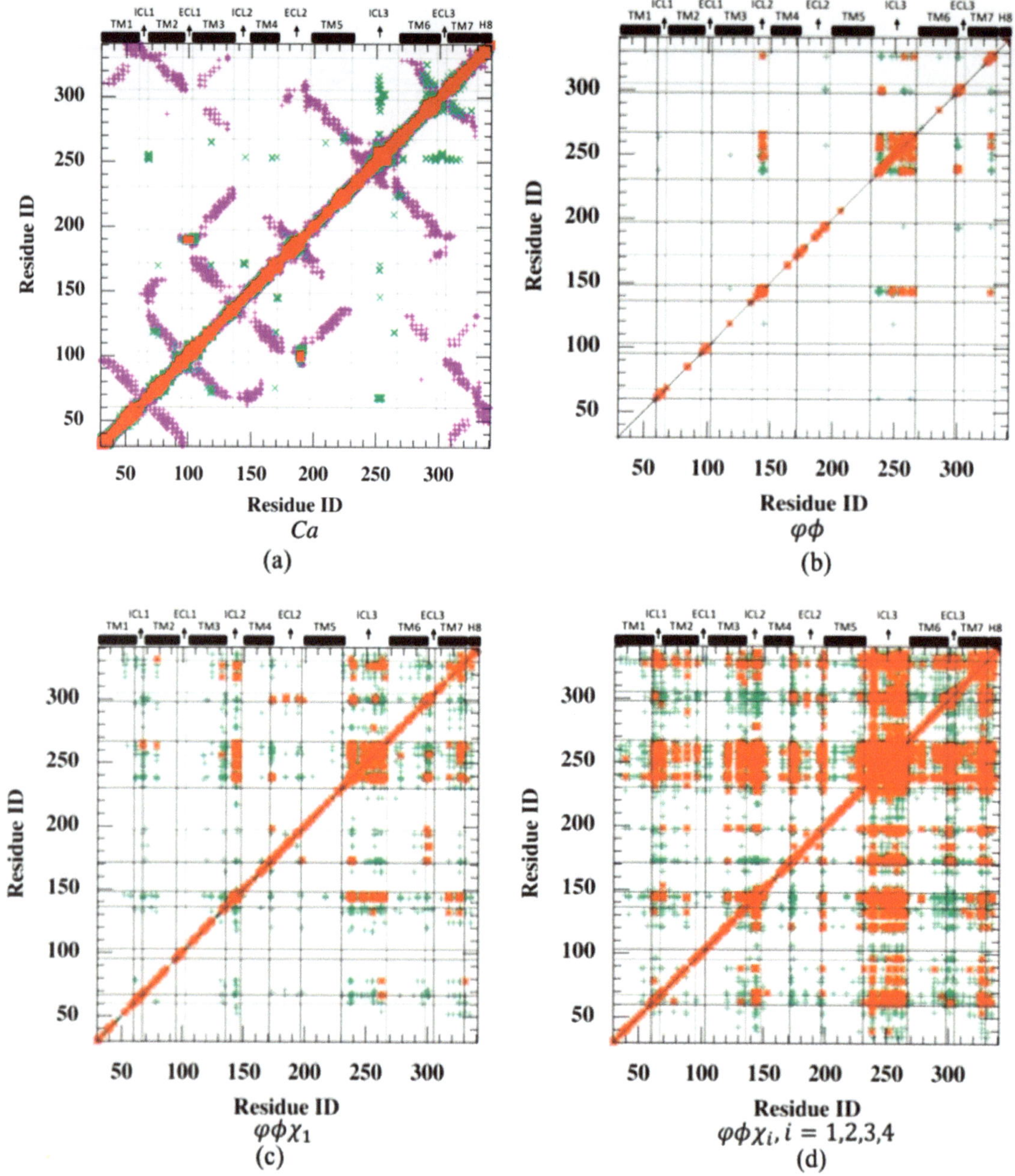

Figure 3. *Cont.*

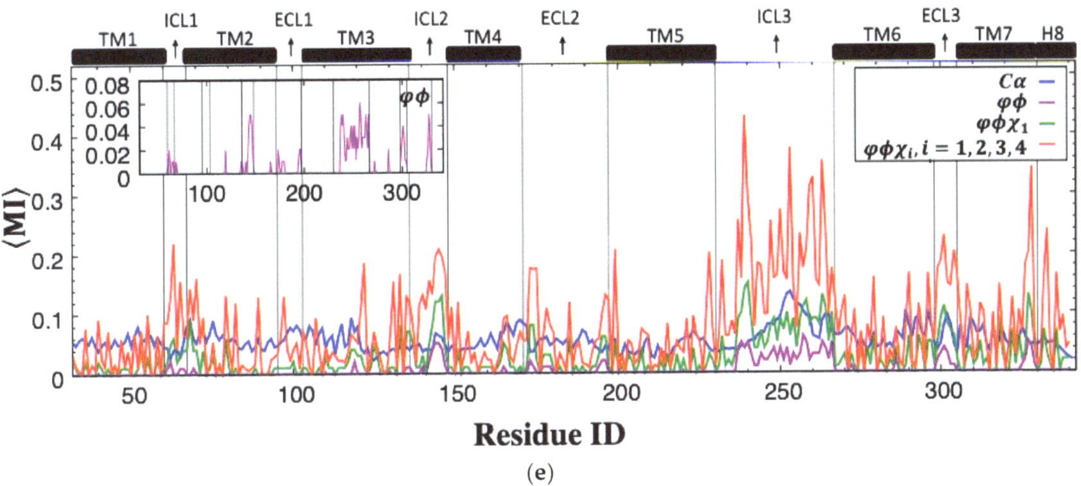

(e)

Figure 3. Residue-pair mutual information (MI) plotted with contact map (threshold: 6.0 Å) for (**a**) backbone C_α, (**b**) backbone torsion angles (ϕ, ψ) (**c**) backbone torsion and first side-chain torsion angles (ϕ, ψ, χ_1), (**d**) backbone torsion and all side-chain torsion angles (ϕ, ψ, χ_i, $i = 1, 2, 3, 4$). (**e**) Mean mutual information per residue averaged over the rest of the receptor. (Green dots: $0.25 < MI \leq 0.5$; red dots: $MI > 0.5$).

The residue-pair MI was next determined for backbone torsion angles ϕ and ψ in each residue. Despite low MI values observed almost everywhere in the receptor (highest MI = 2.59), the relatively high correspondence between ICL3 and ICL2 regions is noticeable (See Figure 3b). Conformational degrees of freedom were mostly dominated by torsion angles in loop regions; thus, the correspondence of loop regions was anticipated. However, the amount of such correspondence appeared to be limited to a few loop regions when only backbone torsion angles were incorporated. Hence, the next attempt was to combine the information of backbone torsion angles with that of the first side-chain torsion angle, χ_1. Maximum MI was slightly increased to 3.72 from 2.59, and the highest MI values were still observed between ICL3 and ICL2 with increased intensity, as depicted in Figure 3c. Additionally, both ICL3 and ICL2 started to share information with the intracellular part of TM7, and moreover, two distant extracellular loop regions, ECL2 and ECL3, displayed some noticeable correspondence with each other.

The total effect of rotational degrees of freedom on MI values can only be disclosed when all possible side-chain torsion angles (χ_i, $i = 1, 2, 3, 4$) were considered together with backbone torsion angles. As illustrated in Figure 3d,e, the increasing trend in MI values between ICL3 and ICL2 was noticeable. Maximum MI reached a value of 6.55. In addition, ICL3 started to share information with the majority of the receptor, including mainly loop regions such as ICL1, ICL2 at the intracellular part, two ends of the extracellular loop ECL2, the entire ECL3, and also the intracellular part of TM7 with its adjacent tail H8. Overall, it is clear that mutual correspondence driven by torsional degrees of freedom mainly existed between loop regions.

It is important to identify residue types most often involved in sharing mutual information, especially among distal ones, as they might point to potential allosteric hub regions. Residue pairs were categorized based on the degree of separation of two residues in the primary sequence, as proximal if 1–4 positions apart and otherwise distal. As illustrated in Figure 4, the highest amount of MI was shared among polar residues, which incorporated *Arg* predominantly. Moreover, two bulky residues *Phe* and *Tyr*, with the highest MI values among hydrophobic residues, also paired with polar *Arg* to a large extent. On the other hand, no significant correspondence was observed among hydrophobic residues. The dominating feature of polar residues in sharing MI can be attributed to their abundance in loop regions, which displayed a noticeably higher amount of MI than transmembrane regions in addition to their higher amount of rotational degrees of freedom. As illustrated in Figure 4c, the frequency of a residue type in loop regions is slightly proportional to its average shared MI. Almost all polar residues indicated with red dots displayed frequency values above 5%, which represents the random occurrence, whereas only three hydrophobic residues, *Gly*, *Phe* and *Leu*, had frequencies above 5%. On the other hand, two polar residues *Ser* and *Thr*, both with hydroxyl groups in their side-chains and frequency values above 7%, displayed low average MI values.

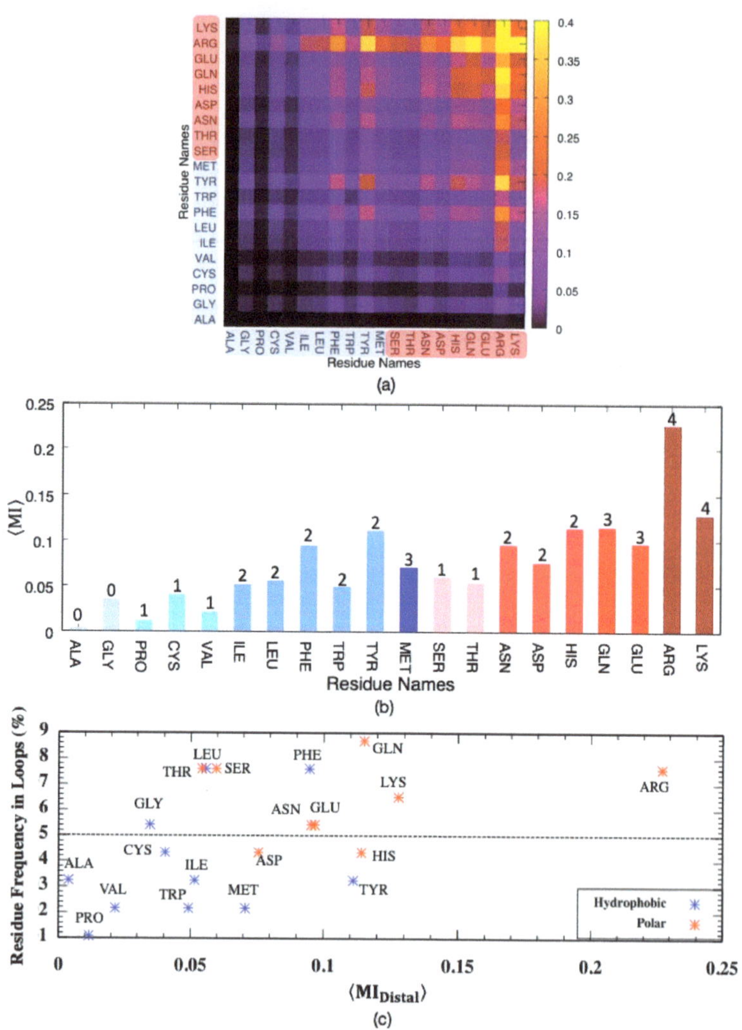

Figure 4. Mutual information for pairs of distal residues classified as either hydrophobic (blue tones) or polar (pink tones) considering all backbone (ϕ, ψ) and side-chain torsion angles (χ_k, $k = 0, 1, 2, 3, 4$), averaged over (**a**) residue pairs, (**b**) residue types with the number of side-chain torsion angles k and (**c**) residue frequency in loop regions versus average MI.

Interestingly, *Arg* and *Lys*, with four rotatable bonds on their positively charged side-chains, displayed slightly different average MI values with respect to each other; with an average MI of 0.23 ± 3.64, *Arg* shared twice as much information as that of *Lys* despite the fact that both residues were found in nearly equal amounts everywhere in the receptor (~7% in loops and ~3% in transmembrane helices, See Supplementary Figure S1). The side-chain of *Arg* has a positively charged guanidium moiety, which favors π-stacking with aromatic rings and has the potential of forming five hydrogen bonds. As the side-chain of *Arg* protrudes from the surface of the protein, it has a crucial role in protein–protein interactions. Hence, it is not surprising to see the highest correspondence between *Arg* and the majority of residues, especially polar ones.

Mutual information is a measure of correspondence between two residues *i* and *j* with respect to their dynamic behaviors, such as positional fluctuations (Cα) or torsional changes happening at the exact same moment. Transfer entropy is another important feature that relates the dynamic states of two residues separated by a time lag parameter τ. In other words, transfer entropy measures the amount of information transferred from residue *i* to residue *j* at a later time. Knowing the state of residue *i*, the state of another residue *j* at a distant site can be predicted if there exists an allosteric communication pathway connecting the two sites. Similar to mutual information analysis, net transfer entropy was first calculated based on the backbone Cα atom's positional fluctuations (see Equations (7) and (8) in Section 2). As depicted in Figure 5a, in one of two MD runs, the intracellular loops ICL3, ICL2, TM3, and TM4 moderately appeared as entropy donor sites, whereas the intracellular part of TM1 was detected as a dominant acceptor site. The second MD trajectory displayed a relatively different distribution profile for donor/acceptor sites; ICL3, together with the extracellular part of TM7, appeared as two dominant entropy donor sites, whereas no major acceptor site was detected (See Figure 5b). The maximum amount of information transferred was observed as 16.32 in Run#1 and 18.15 in Run#2. These results clearly indicate alternative communication pathways that can be established via positional fluctuations of Cα atoms for the same system in two separate runs.

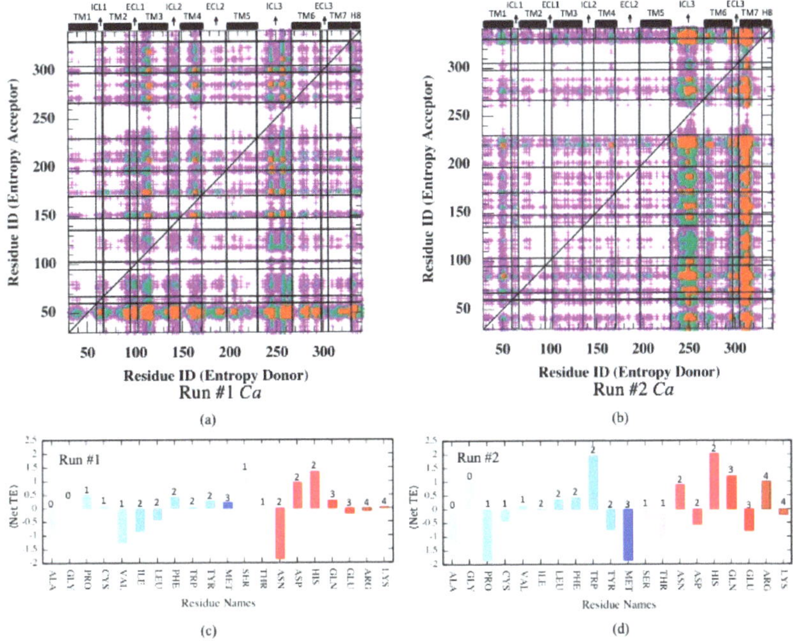

Figure 5. (**a,b**) Net transfer entropy from residue *i* to residue *j* (See Equation (8)) based on Cα fluctuations for two independent MD runs. (**c,d**) Net transfer entropy averaged for each residue type categorized as either hydrophobic (blue tones) or polar (pink tones) and the number of side-chain torsion angles *k* indicated at the top of each bar. Color code for (**a,b**): no display for $\langle netTE \rangle \leq 4$; magenta for $4 < \langle netTE \rangle \leq 8$, green for $8 < \langle netTE \rangle \leq 12$ and red for $12 > \langle netTE \rangle$.

Furthermore, net transfer entropy was averaged for each residue type categorized as either hydrophobic or polar, as depicted in Figure 5c,d, where positive and negative $\langle netTE \rangle$ values correspond to entropy donor and acceptor residues, respectively. Clearly, no correspondence was detected between the number of side-chain torsion angles and the net entropy in both runs. In addition, there is no clear tendency for polar or hydrophobic residues to display either entropy donor or acceptor features. Furthermore, two runs displayed two completely different donor/acceptor profiles with respect to residue types. This is especially noticeable in $\langle netTE \rangle$ maps illustrated for all residue pairs, such as entropy donor versus entropy acceptor depicted in Supplementary Figure S2. Several residues displayed opposite features, e.g., entropy donor in one run and acceptor in the second run, such as *Pro, Cys, Val, Ile, Leu, Tyr, Met, Ser, Thr, Asn, Asp, Glu,* and *Lys*. Apparently, fluctuations in Cα displacements were not driven by residue type, which incorporates the information of both polarity/hydrophobicity and the number of degrees of freedom.

Transfer entropy was next determined using backbone torsion angles, (φ, ψ). As depicted in Figure 6a, ICL3 appeared as the only source of entropy donor to a few isolated acceptor regions detected on mostly loops such as ICL1, ECL1, ECL2, ECL3 and the intracellular part of TM7 adjacent to segment H8. Maximum TE values were determined as 10.86 and 11.54 for two runs, which are well below Cα-based TE values. Next, the first side-chain torsion angle (χ_1) was considered together with two backbone angles for identifying the information of the rotameric state transferred from one residue to another in the receptor. As illustrated in Figure 6b, the same loop regions still appeared as entropy donor sites with an increased intensity dominating the future fluctuations of torsional angles everywhere in the receptor. In addition, maximum TE values reached 26.46 and 31.54 in two runs.

Unfortunately, the addition of more than one side-chain torsion angle made the computation intractable due to triple joint probability calculations (See Equation (7)), as it roughly required a memory space of $523,792,501,128$ bytes $(= 3^{18} \times N_{Arg} \times N_{Lys} \times 8(\text{bytes}/\text{ArrayCell}))$ only for calculating the $p\left(\Delta R_j(t), \Delta R_j(t-\tau), \Delta R_i(t-\tau)\right)$ parameter of the transfer entropy equation between all *Arg* and *Lys* pairs, which exceeded the maximum amount available for today's computer technology. However, the same analysis was conducted for all possible side-chain torsion angles χ_k, $k = 1, 2, 3, 4$ only. As anticipated, information transferred from one region to another site increased significantly, with a maximum TE value reaching 61.34 for Run #1 and 63.38 for Run #2 (see Figure 6c). All intra- and extracellular loops that extended slightly towards the neighboring helices were detected as important entropy donor sites. These results clearly represent that the conformational states of the side-chains at loop regions extensively dominated the future conformational states of side-chain torsion angles everywhere in the receptor.

Finally, the net transfer entropy was further decomposed and replotted for each of the 20 residue types, as depicted in Figure 7, using a bar plot to display the average net transfer entropy where the entropy source (donor) and sink (acceptor) residues can be identified by their positive and negative values, respectively. Corresponding plots that display net transfer entropy for a pair of residue types such as entropy donor versus entropy acceptor are provided in Supplementary Figure S3. Further categorization of residues as hydrophobic and polar clearly demonstrated the dominancy of polar residues as entropy donors, whereas hydrophobic ones were most often identified as entropy acceptors. In the case of backbone rotation angles only, *Trp* appeared as the strongest entropy acceptor site in both MD runs, whereas *His* displayed the highest positive average net transfer entropy (See Figure 7a). Exceptionally, *Gly* residue with no side-chain atoms appeared as a strong entropy donor site.

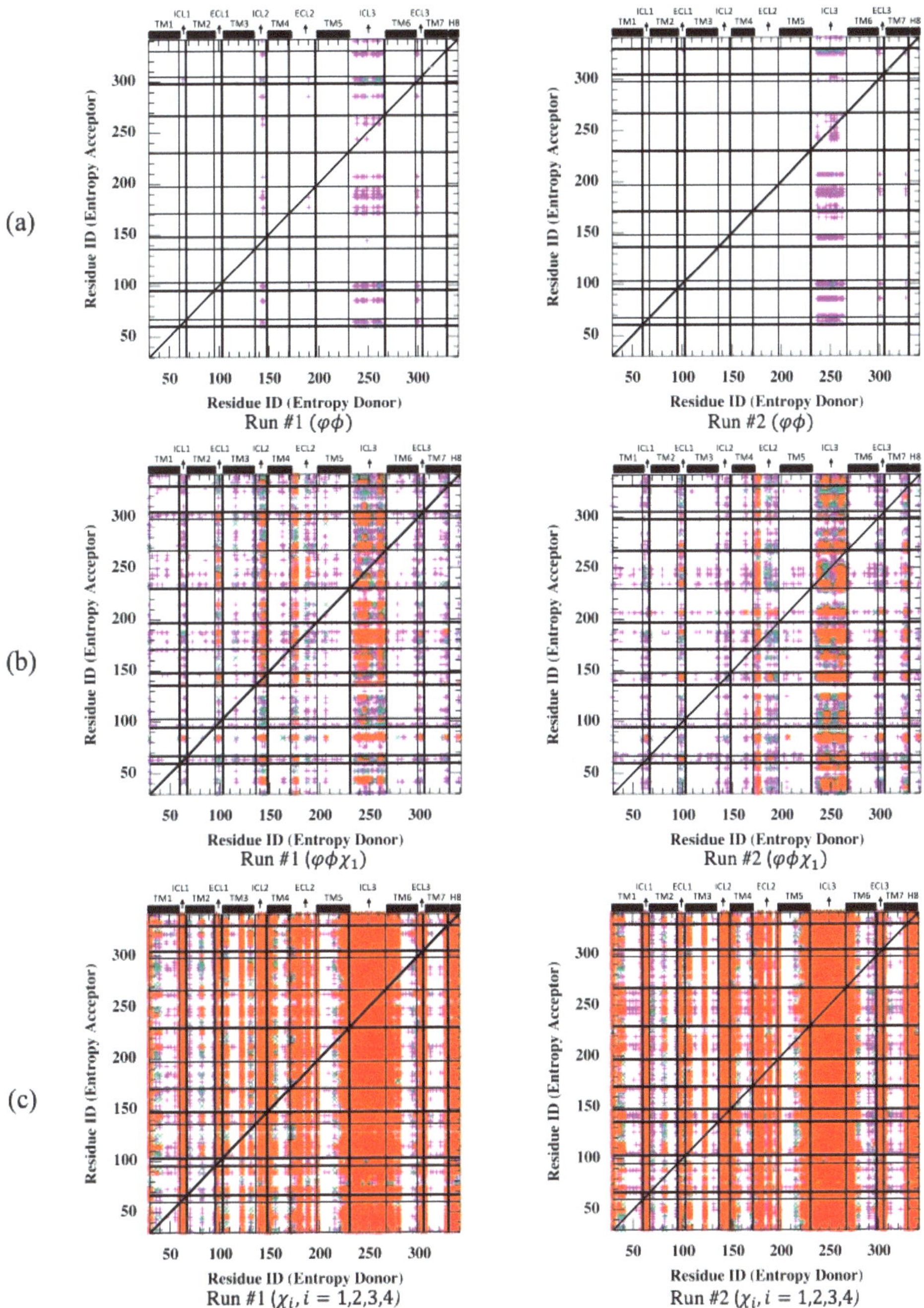

Figure 6. Net transfer entropy from residue i to residue j (see Equation (8)) for two independent MD runs using (**a**) backbone torsion angles, (ϕ, ψ), (**b**) backbone torsion angles, (ϕ, ψ) and the first side-chain torsion angle (χ_1) and (**c**) all possible side-chain torsion angles χ_k, $k = 1, 2, 3, 4$. Color code for (**a**,**b**): no display for $\langle netTE \rangle \leq 4$; magenta for $4 < \langle netTE \rangle \leq 8$, green for $8 < \langle netTE \rangle \leq 12$ and red for $12 > \langle netTE \rangle$.

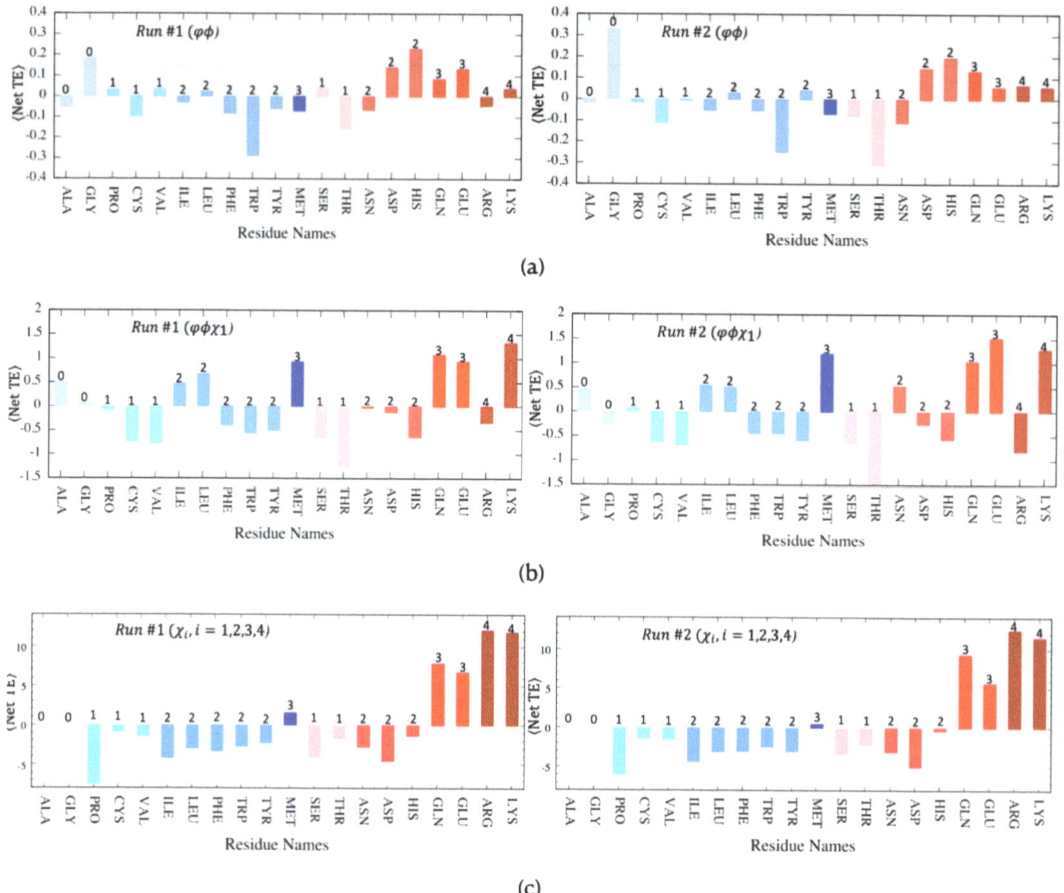

Figure 7. Net transfer entropy averaged for each residue type categorized as either hydrophobic (blue tones) or polar (pink tones). The number of side-chain torsion angles *k* indicated at the top of each bar is determined for both MD runs using (**a**) backbone torsion angles, (ϕ, ψ), (**b**) backbone torsion angles, (ϕ, ψ) and the first side-chain torsion angle (χ_1) and (**c**) all possible side-chain torsion angles χ_k, $k = 1, 2, 3, 4$.

Noticeably, the close correspondence between polarity/hydrophobicity and donor/acceptor features was the strongest when all side-chain torsion angles were considered in transfer entropy calculations (See Figure 7c). Most hydrophobic residues, except *Met*, which is mostly located at the protein's core region, displayed strong entropy acceptor characteristics, especially *Pro*, with the lowest average net transfer entropy value of -7.38 calculated so far. Furthermore, on the polar side, *Arg* and *Lys*, with a total of four side-chain torsion angles, displayed the highest entropy values exceeding $+10$. Upon incorporating the first side-chain torsion angle (χ_1) along with two backbone torsion angles, the profile changed slightly, yet the dominance of polar residues as entropy donor sites persisted (See Figure 7b). Three polar residues, *Lys, Glu, and Gln*, displayed the highest positive transfer entropy values in both runs. Interestingly, the two polar *Ser, Thr*, and the hydrophobic *Tyr*, which all contain a hydroxyl group in their side-chain, displayed entropy sink (acceptor) features.

4. Conclusions

Two independent 1.5 μs long MD simulations were conducted on the apo form of the active state of human β$_2$AR in a complex with a Gs protein. Throughout both trajectories, the active state of the receptor was well preserved with the characteristic tilt in transmembrane helix 6 and ICL3 towards the lipid bilayer to give Gs full access to the binding cavity at the intracellular part. On the extracellular part, since no ligand was attached at the orthosteric site, a minor expansion was observed because of the slightly distancing motion of TM5 from TM3. However, this slight conformational shift at the extracellular part did not cause any allosteric interference in the intracellular region.

Distant regions fluctuating in the correspondence are critical as they might point to potential sites along the allosteric pathway. In this study, we attempted to use several metrics for that purpose. First, residue-pair cross-correlations were calculated for α-Carbon atomic fluctuations from average positions. Distant and correlated regions were mainly observed within the last three transmembrane helices (TM5, TM6, and TM7), including the longest loop region ICL3 and the small extension H8 adjacent to TM7. Moreover, TM6 and TM7 fluctuated in opposite directions with TM1. As cross-correlation ignores the correlated motions in orthogonal directions, the mutual information metric was next used to identify all possible distant sites in correspondence that might be critical for allosteric signaling. First, only α-Carbon atomic fluctuations were considered. However, not much correspondence was detected in the receptor except between ICL3 and the distant extracellular parts of TM6 and TM7. The next step was to replace α-Carbon fluctuations with rotameric states of backbone torsion angles ϕ and ψ in each residue when formulating the mutual correspondence. A considerable change was observed in the profile where ICL3's rotameric states fluctuated in concert and with respect to ICL2. Incorporating side-chain torsion angles further increased the mutual information transferred between ICL3 and ICL2. In addition, ICL3 started to share information with all the other loop regions, including some limited portions of transmembrane helices, TM3, TM6, and TM7.

When mutual information was further decomposed based on types of residue pairs, polar ones, especially *Arg*, were identified as the dominating group sharing the highest correspondence with other polar residues. Hydrophobic residues shared the least amount of mutual information, except *Tyr* and *Phe*, which paired with polar *Arg*. The lowest amount of MI was observed among hydrophobic residues. The dominating feature of polar residues was attributed to their higher abundance in loop regions where the highest mutual information was detected. However, despite its low abundance in loop regions, hydrophobic *Tyr* with two side-chain torsion angles had average mutual information of 0.11, which was comparable with that of other polar residues with two side-chain torsion angles.

Transfer entropy, which is another metric in information theory, relates two states at different times. If the state of residue *j* in the future time can be predicted knowing the state of residue *i* at the present time, then two sites might communicate with each other as part of an allosteric signaling network. First, transfer entropy was determined for the information about positional fluctuations (Cα). Different profiles were observed in two independent MD runs. In one run, ICL3, ICL2, TM3, and TM4 moderately appeared as entropy donor sites, whereas the intracellular part of TM1 was detected as a dominant acceptor site. In the second run, ICL3, together with the extracellular part of TM7, appeared as two dominant entropy donor sites, whereas no major acceptor site was detected. Clearly, there is no unique communication pathway for backbone Cα displacements. When the information type was replaced by the fluctuation in the rotameric states of backbone torsion angles (φ, ψ), a completely different profile of communication network appeared and persisted in both runs; ICL3 was the only source of entropy donor to a few isolated acceptor regions detected mostly on loops such as ICL1, ECL1, ECL2, ECL3 and the intracellular part of TM7 adjacent to segment H8. The intensity of transferred information was relatively weak, 10.86 and 11.54 for two runs, which were well below Cα-based TE values (16.32 and 18.15). Then, the rotameric states of the first side-chain torsion angle (χ_1) were combined with those of backbone torsion angles. As anticipated, the intensity of transferred information noticeably increased with a maximum value of 26.46 and 31.24 in two runs. Due to computational limitations, the addition of another side-chain torsion angle was not achievable, yet the increasing trend in transfer entropy was predictable. When only the rotameric states of side-chain torsion angles were used, transfer entropy significantly increased to its highest values (61.34 and 63.38), yet the distribution profile among regions was preserved, i.e., the fluctuations of torsion angles in the loop regions drove the future fluctuations of rotameric states everywhere in the receptor. This result clearly elucidates an important aspect of all GPCRs where both extra- and intracellular loops protruding from the transmembrane bilayer play a major role in the functional regulation. Thus, loop regions can be essential targets for the design of allosteric drug molecules with fewer side effects.

Supplementary Materials: The following supporting information can be downloaded at: https://www.mdpi.com/article/10.3390/app12178530/s1, LIST OF SUPPLEMENTARY FIGURES: Figure S1. Residue frequency (%) of each residue type in loop and transmembrane (TM) regions. Figure S2. (a,b) Average net transfer entropy based on Cα fluctuations for residue pairs classified as either hydrophobic (blue) or polar (pink) for two MD runs. Figure S3. (a,b) Average net transfer entropy for residue pairs classified as either hydrophobic (blue) or polar (pink) for two MD runs based on rotameric states of (a) backbone torsion angles, (ϕ, ψ), (b) backbone torsion angles, (ϕ, ψ) and the first side-chain torsion angle (χ_1) and (c) all possible side-chain torsion angles χ_k, $k = 1, 2, 3, 4$.

Author Contributions: Conceptualization, E.D.A.; methodology, N.S. and E.D.A.; software, N.S. and E.D.A.; validation N.S. and E.D.A.; formal analysis, N.S. and E.D.A.; investigation, N.S.; resources, E.D.A.; data curation, N.S. and E.D.A.; writing—original draft preparation, E.D.A.; writing—review and editing, E.D.A.; visualization, N.S.; supervision, E.D.A.; project administration, E.D.A.; funding acquisition, N.S. All authors have read and agreed to the published version of the manuscript.

Funding: This research received no external funding.

Institutional Review Board Statement: Not applicable.

Informed Consent Statement: Not applicable.

Data Availability Statement: The data presented in this study are available on request from the corresponding author. The data are not publicly available due to the privacy of the source code.

Acknowledgments: N.S. acknowledges The Scientific and Technological Research Council of Turkey (TUBITAK) for her 2211/C National Ph.D. Scholarship. TOC Graphics were created with BioRender.com.

Conflicts of Interest: The authors declare no conflict of interest.

References

1. Gunasekaran, K.; Ma, B.; Nussinov, R. Is allostery an intrinsic property of all dynamic proteins? *Proteins Struct. Funct. Bioinf.* **2004**, *57*, 433–443. [CrossRef] [PubMed]
2. Sun, J.H.; O'Boyle, D.R.; Fridell, R.A.; Langley, D.R.; Wang, C.; Roberts, S.B.; Nower, P.; Johnson, B.M.; Moulin, F.; Nophsker, M.J.; et al. Resensitizing daclatasvir-resistant hepatitis C variants by allosteric modulation of NS5A. *Nature* **2015**, *527*, 245–248. [CrossRef] [PubMed]
3. Hayouka, Z.; Rosenbluh, J.; Levin, A.; Loya, S.; Lebendiker, M.; Veprintsev, D.; Kotler, M.; Hizi, A.; Loyter, A.; Friedler, A. Inhibiting HIV-1 integrase by shifting its oligomerization equilibrium. *Proc. Natl. Acad. Sci. USA* **2007**, *104*, 8316–8321. [CrossRef] [PubMed]
4. Mikulska-Ruminska, K.; Shrivastava, I.H.; Krieger, J.M.; Zhang, S.; Li, H.; Bayir, H.; Wenzel, S.E.; VanDemark, A.P.; Kagan, V.E.; Bahar, I. Characterization of differential dynamics, specificity, and allostery of lipoxygenase family members. *J. Chem. Inf. Model.* **2019**, *59*, 2496–2508. [CrossRef] [PubMed]
5. Cooper, A.; Dryden, D.T.F. Allostery without conformational change—A plausible model. *Eur. Biophys. J.* **1984**, *11*, 103–109. [CrossRef] [PubMed]
6. Tsai, C.J.; Nussinov, R. A unified view of "how allostery works". *PLoS Comput. Biol.* **2004**, *10*, e1003394. [CrossRef] [PubMed]
7. Motlagh, H.N.; Wrabl, J.O.; Li, J.; Hilser, V.J. The ensemble nature of allostery. *Nature* **2014**, *508*, 331–339. [CrossRef]
8. Popovych, N.; Sun, S.; Ebright, R.H.; Kalodimos, C.G. Dynamically driven protein allostery. *Nat. Struct. Mol. Biol.* **2006**, *13*, 831–838. [CrossRef]
9. Guarnera, E.; Berezovsky, I.N. Allosteric drugs and mutations: Chances, challenges, and necessity. *Curr. Opin. Struct. Biol.* **2020**, *62*, 149–157. [CrossRef]
10. Nussinov, R.; Tsai, C.J. Allostery in disease and in drug discovery. *Cell* **2013**, *153*, 293–305. [CrossRef]

11. Shrivastava, I.H.; Liu, C.; Dutta, A.; Bakan, A.; Bahar, I. *Allostery as Structure-Encoded Collective Dynamics: Significance in Drug Design*; Structural Biology in Drug Discovery: Methods, Techniques, and Practices; Renaud, J.P., Ed.; John Wiley & Sons: New York, NY, USA, 2020; Chapter 6; pp. 125–141.

12. Ayyildiz, M.; Celiker, S.; Ozhelvaci, F.; Akten, E.D. Identification of alternative allosteric sites in glycolytic enzymes for potential use as species-specific drug targets. *Front. Mol. Biosci.* **2020**, *7*, 88. [CrossRef] [PubMed]

13. Celebi, M.; Inan, T.; Kurkcuoglu, O.; Akten, E.D. Potential allosteric sites captured in glycolytic enzymes via residue-based network models: Phosphofructokinase, glyceraldehyde-3-phosphate dehydrogenase and pyruvate kinase. *Biophys. Chem.* **2022**, *280*, 106701. [CrossRef] [PubMed]

14. Kurochkin, I.V.; Guarnera, E.; Wong, J.H.; Eisenhaber, F.; Berezovsky, I.N. Toward Allosterically Increased Catalytic Activity of Insulin- Degrading Enzyme against Amyloid Peptides. *Biochemistry* **2017**, *56*, 228–239. [CrossRef] [PubMed]

15. Wang, J.; Jain, A.; McDonald, L.R.; Gambogi, C.; Lee, A.L.; Dokholyan, N.V. Mapping allosteric communications within individual proteins. *Nat. Commun.* **2020**, *11*, 3862. [CrossRef]

16. Amor, B.R.C.; Schaub, M.T.; Yaliraki, S.N.; Barahona, M. Prediction of allosteric sites and mediating interactions through bond-to-bond propensities. *Nat. Commun.* **2016**, *7*, 12477. [CrossRef] [PubMed]

17. Kaya, C.; Armutlulu, A.; Ekesan, S.; Haliloglu, T. MCPath: Monte Carlo path generation approach to predict likely allosteric pathways and functional residues. *Nucleic Acids Res.* **2013**, *41*, W249–W255. [CrossRef]

18. Schreiber, T. Measuring information transfer. *Phys. Rev. Lett.* **2000**, *85*, 461–464. [CrossRef]

19. Kamberaj, H.; van der Vaart, A. Extracting the Causality of Correlated Motions from Molecular Dynamics Simulations. *Biophys. J.* **2009**, *97*, 1747–1755. [CrossRef]

20. Barr, D.; Oashi, T.; Burkhard, K.; Lucius, S.; Samadani, R.; Zhang, J.; Shapiro, P.; MacKerell, A.D., Jr.; van der Vaart, A. Importance of domain closure for the autoactivation of ERK2. *Biochemistry* **2011**, *50*, 8038–8048. [CrossRef]

21. Corrada, D.; Morra, G.; Colombo, G. Investigating allostery in molecular recognition: Insights from a computational study of multiple anti- body–antigen complexes. *J. Phys. Chem. B* **2013**, *117*, 535–552. [CrossRef]

22. Hacisuleyman, A.; Erman, B. Causality, transfer entropy, and allosteric communication landscapes in proteins with harmonic interactions. *Proteins Struct. Funct. Bioinf.* **2017**, *85*, 1056–1064. [CrossRef] [PubMed]

23. Hacisuleyman, A.; Erman, B. Entropy transfer between residue pairs and allostery in proteins: Quantifying allosteric communication in ubiquitin. *PLoS Comput. Biol.* **2017**, *13*, e1005319. [CrossRef] [PubMed]

24. DuBay, K.H.; Geissler, P.L. Calculation of proteins' total side-chain torsional entropy and its influence on protein-ligand interactions. *J. Mol. Biol.* **2009**, *391*, 484–497. [CrossRef]

25. Marlow, M.S.; Dogan, J.; Frederick, K.K.; Valentine, K.G.; Wand, A.J. The role of conformational entropy in molecular recognition by calmodulin. *Nat. Chem. Biol.* **2010**, *6*, 352–358. [CrossRef] [PubMed]

26. Millet, O.; Mittermaier, A.; Baker, D.; Kay, L.E. The effects of mutations on motions of side-chains in protein l studied by 2 h nmr dynamics and scalar couplings. *J. Mol. Biol.* **2003**, *329*, 551–563. [CrossRef]

27. DuBay, K.H.; Bothma, J.P.; Geissler, P.L. Long-Range Intra-Protein Communication Can Be Transmitted by Correlated Side-Chain Fluctuations Alone. *PLoS Comput. Biol.* **2011**, *7*, e1002168. [CrossRef]

28. Rasmussen, S.G.F.; Choi, H.J.; Fung, J.J.; Pardon, E.; Casarosa, P.; Chae, P.S.; Kobilka, B.K. Structure of a nanobody-stabilized active state of the β2 adrenoceptor. *Nature* **2011**, *469*, 175–180. [CrossRef]

29. Eswar, N.; Webb, B.; Marti-Renom, M.A.; Madhusudhan, M.S.; Eramian, D.; Shen, M.; Pieper, U.; Sali, A. Comparative protein structure modeling using Modeller. *Curr. Protoc. Bioinform.* **2006**, *15*, 5–6. [CrossRef]

30. Humphrey, W.; Dalke, A.; Schulten, K. VMD: Visual molecular dynamics. *J. Mol. Graph.* **1996**, *14*, 33–38. [CrossRef]

31. Best, R.B.; Zhu, X.; Shim, J.; Lopes, P.E.M.; Mittal, J.; Feig, M.; MacKerell, A.D., Jr. Optimization of the additive CHARMM all-atom protein force field targeting improved sampling of the backbone phi, psi and side-chain chi1 and chi2 dihedral angles. *J. Chem. Theory Comput.* **2012**, *8*, 3257–3273. [CrossRef]

32. Feller, S.E.; Zhang, Y.H.; Pastor, R.W. Computer-simulation of liquid/liquid interfaces 2. Surface-tension area dependence of a bilayer and monolayer. *J. Chem. Phys.* **1995**, *103*, 10267–10276. [CrossRef]

33. Phillips, J.C.; Braun, R.; Wang, W.; Gumbart, J.; Tajkhorshid, E.; Villa, E.; Chipot, C.; Skeel, R.D.; Kalé, L.; Schulten, K. Scalable molecular dynamics with NAMD. *J. Comput. Chem.* **2005**, *26*, 1781–1802. [CrossRef] [PubMed]

34. Petrache, H.I.; Dodd, S.W.; Brown, M.F. Area per lipid and acyl length distributions in fluid phosphatidylcholines determined by (2)H NMR spectroscopy. *Biophys. J.* **2000**, *79*, 3172–3319. [CrossRef]

35. Kučerka, N.; Nieh, M.P.; Katsaras, J. Fluid phase lipid areas and bilayer thicknesses of commonly used phosphatidylcholines as a function of temperature. *Biochim. Biophys. Acta-Biomembr.* **2011**, *1808*, 2761–2771. [CrossRef] [PubMed]

36. Ponder, J.W.; Richards, F.M. Tertiary templates for proteins. Use of packing criteria in the enumeration of allowed sequences for different structural classes. *J. Mol. Biol.* **1987**, *193*, 775–791. [CrossRef]

37. Ozgur, C.; Doruker, P.; Akten, E.D. Investigation of allosteric coupling in human β2-adrenergic receptor in the presence of intracellular loop 3. *BMC Struct. Biol.* **2016**, *16*, 9. [CrossRef]

38. Nygaard, R.; Zou, Y.; Dror, R.O.; Mildorf, T.J.; Arlow, D.H.; Manglik, A.; Pan, A.C.; Liu, C.W.; Fung, J.J.; Bokoch, M.P.; et al. The dynamic process of β2-adrenergic receptor activation. *Cell* **2013**, *152*, 532–542. [CrossRef] [PubMed]

39. Ozcan, O.; Uyar, A.; Doruker, P.; Akten, E.D. Effect of intracellular loop 3 on intrinsic dynamics of human β2-adrenergic receptor. *BMC Struct. Biol.* **2013**, *13*, 29. [CrossRef]

applied sciences

Perspective

Are Protein Shape-Encoded Lowest-Frequency Motions a Key Phenotype Selected by Evolution?

Laura Orellana

Protein Dynamics and Mutation Lab, Department of Oncology-Pathology, Karolinska Institute, Solnavägen 9, 171 65 Solna, Sweden; laura.orellana@ki.se

Abstract: At the very deepest molecular level, the mechanisms of life depend on the operation of proteins, the so-called "workhorses" of the cell. Proteins are nanoscale machines that transform energy into useful cellular work, such as ion or nutrient transport, information processing, or energy transformation. Behind every biological task, there is a nanometer-sized molecule whose shape and intrinsic motions, binding, and sensing properties have been evolutionarily polished for billions of years. With the emergence of structural biology, the most crucial property of biomolecules was thought to be their 3D shape, but how this relates to function was unclear. During the past years, Elastic Network Models have revealed that protein shape, motion and function are deeply intertwined, so that each structure displays robustly shape-encoded functional movements that can be extraordinarily conserved across the tree of life. Here, we briefly review the growing literature exploring the interplay between sequence evolution, protein shape, intrinsic motions and function, and highlight examples from our research in which fundamental movements are conserved from bacteria to mammals or selected by cancer cells to modulate function.

Keywords: protein dynamics; evolution; intrinsic motions; elastic network models

1. From the Structure–Function Paradigm to Structure–Motion–Function

Over 60 years ago, Anfinsen's postulate that "the native secondary and tertiary structures are contained in the amino acid sequence itself" [1] laid out the foundations of the central dogma of structural biology, i.e., that the sequence of a protein contains the information required to adopt a defined 3D-structure and, hence, function (see historical overview in [2]). This so-called structure–function paradigm was formulated during the time when biomolecular crystallography was flourishing. According to Martin Karplus, X-ray crystallography created "the misconception … that the atoms in a protein are fixed in position" [3]. This view is also shared by cryo-EM pioneer Joachim Frank, who wrote that "the idea of "a" molecular structure has been largely created by X-ray crystallographic practice" [4]. As a consequence, a static view of proteins, in which one sequence folds into a unique "native conformation" responsible for function, became prevalent. Nevertheless, an alternative, dynamic view of proteins as an ensemble of conformations, more akin to the principles of physics, had been proposed long before by Pauling, Landsteiner, and others in the 1930s [5]. Fast forward in time to our days, and this early dynamic vision appears prescient. As our technology to capture proteins in action evolved (NMR, cryo-EM, etc.), it became clearer every day that proteins do not fold into a single static "native" structure, but are rather dynamic machines in continuous motion that explore complex and rugged energy landscapes [6], transitioning between multiple meta-stable minima. Such transitions encompass a wide hierarchy of time and length scales—from picosecond atomic fluctuations to microsecond or millisecond allosteric changes or breathing motions—and, importantly, are instrumental for proteins to sense and respond to environmental signals like ions or ligands [6–8].

Protein motions not only mediate or execute biological work—channel gating, ion pumping, transport, etc.—but also reshape interactions with other partners. Therefore, they

Citation: Orellana, L. Are Protein Shape-Encoded Lowest-Frequency Motions a Key Phenotype Selected by Evolution? *Appl. Sci.* **2023**, *13*, 6756. https://doi.org/10.3390/app13116756

Academic Editors: Robert Jernigan and Domenico Scaramozzino

Received: 7 February 2023
Revised: 17 May 2023
Accepted: 19 May 2023
Published: 1 June 2023

are central for molecular recognition [9–11], no matter whether it involves conformational selection or induced fit [12,13]. Even eminently local processes such as enzyme catalysis can involve dynamic changes such as side chain fluctuations or the unfolding of binding sites [14–16]. For intrinsically disordered proteins, flexibility is so extreme that the classical concept of a discrete number of well-defined native 3D shapes or conformers becomes almost meaningless; they can only be statistically described as ensembles of interconverting conformations [17,18]. Nevertheless, a majority of proteins fall in the middle ground between perfect rigidity and chaotic disorder, a boundary where discrete rigid domains or subunits exquisitely rearrange in response to signals. Cooperative motions, allosteric propagation, and large-scale conformational changes spontaneously emerge from this frontier of harnessed flexibility to create function, as pioneering work by Dorothee Kern showed [16].

Back in 1987, Elber and Karplus first noted the similarity of MD fluctuations with evolutionary changes across the globin family [19], inaugurating a fruitful line of evolutionary and structural dynamics comparisons to this day. Since then, structural data have grown exponentially, and Elastic Network Models (ENMs) [20–23] have revealed that such fluctuations are largely defined by molecular shape and determine functional motions. Overall, this has led to a new structure–motion–function dogma, where molecular shape determines intrinsic motions, and motions make function, a concept increasingly supported via cryo-EM ensembles [24,25]. Therefore, it is time to ask: if molecular motions mediate function, are they maybe a key object of evolutionary selection? Here, we briefly review evidence from structural biology and ENMs research, that points to shape-encoded motions as an essential matter for evolution.

2. ENMs Overview and the Surprising Accuracy of Shape-Encoded Harmonic Motions

A central problem in the study of protein dynamics has always been the difficulty of capturing motion, i.e., fully sampling conformational spaces. Protein flexibility is challenging to trap, describe, and predict, both experimentally and computationally [26]. Despite advances in hardware and algorithm parallelization, fully atomistic Molecular Dynamics (MD) simulations are still only feasible for ns–μs timescales and middle-sized proteins. To gain insight into the mechanisms of bigger sub-mesoscopic systems or the slow large-scale transitions associated with biological function, the physical description needs to change accordingly to lower-resolution Coarse-Grained (CG) models. Among the plethora of CG methods to model the dynamics of proteins, ENMs stand out as possibly the most simple and powerful, considering the balance between their minimal computational cost and striking predictive power. ENMs can be described as the CG flavor of Normal Mode Analysis (NMA), a classical mechanics technique used since the 1940s–1950s to analyze the vibrational spectra of simple molecules [27,28]. Soon after the first MD simulations, in 1982–1983 [29–33], NMA was applied for the first time to proteins to gain insight into their near-equilibrium dynamics. Instead of numerically solving Newton's equations as MD does, NMA assumes the harmonicity of the system around an energy minimum and, thus, through diagonalization of the mass-weighted Hessian matrix, allows the computation of a unique analytical solution, i.e., a set of linearly independent Normal Nodes (NMs) (see details in [21,34]). NMs are a series of eigenvectors (v_i) ordered by their eigenvalues or frequencies (λ_i), that describe the natural motions of the system. Importantly, the first 5–10 ones, the so-called lowest frequency, "soft" or "slow" modes, capture the largest amplitude, more collective, and energetically "easiest" movements, which usually coincide with the experimentally and biologically relevant ones, as we will discuss below.

Despite its simplicity versus MD, NMA was still computationally heavy for large systems, as it required energy minimization and significant memory resources for matrix diagonalization. Inspired by early "random networks" and "beads-and-springs" polymer models developed by Flory and Rouse [35,36], ENMs took the simplification of NMA one step further, replacing detailed physical force fields with a minimalist representation of proteins as networks of residue nodes connected with elastic springs, devoid of chemical or sequence information. Moreover, the system was assumed to be already at a minimum, skipping energy minimization. The first ENM [37] was still an all-atom model but with a simple pairwise Hookean potential: the native structure was defined as the minimum, and detailed interactions were replaced with a squared potential and a uniform constant within a cutoff. Shortly after, Bahar's one-dimensional Gaussian Network Model (GNM) [38] introduced the coarse-graining of structures to the $C\alpha$ trace, and finally, the Anisotropic Network Model (ANM) [39] combined Tirion's 3D-model with GNM coarse-graining, becoming the basis for most ENM methods nowadays [22,40]. The similarity of the motions described using coarse-grained ENMs with the atomistic Tirion's model, and of Tirion's with classical NMA based on accurate molecular potentials, was initially puzzling. How can such minimal one- or two-parameter models reproduce the vibrational properties of a complex macromolecule? The answer lies in the fact that soft modes involve coherent motions of large groups of atoms, and thus are mostly defined by the overall mass/domain architecture. For that matter, CG and atomistic mappings are nearly equivalent.

ENM–NMA can have apparent simplicity—with "toy" ad hoc force fields and the naïve assumption that structures are in an energy minimum—but it is often unsurpassed in the prediction of experimentally observed large-scale conformational changes (Figure 1, center). There have been endless studies comparing ENMs with functional transitions between bound/unbound, active/inactive and open/closed pairs derived from X-ray conformers, NMR ensembles, etc., which show that the lowest-frequency modes are indeed both biologically and functionally relevant [41–44] and can unravel complex allosteric mechanisms [45], even for subtle transitions such as those seen in GPCRs [46–48]. Protein conformational changes often involve large rigid-body motions, e.g., domain swapping, hinge-bending, or shear movements, which are strikingly well described via a small number of ENM modes [49–51]. An early study on the first database of molecular motions, MolMov [52], determined that 95% of experimentally observed transitions can be described using just a couple of soft ENM modes. Further benchmark studies have confirmed that large-scale motions also coincide with the collective modes extracted from MD simulations or experimental ensembles [53–57] via Principal Components Analysis (PCA, see [58–60]). Systematic comparison with MD of representative meta-folds in the MODEL database as well as with experimental data [61,62] confirmed that ENMs are extremely robust to spring definitions and perform exceedingly well in predicting large-scale transitions, occasionally surpassing MD simulations.

Nevertheless, as often happens with CG models, a major weakness of ENMs is the lack of a consistent and universal consensus on force-field parameterization, i.e., the functions used to determine the "springs" connecting different residues or "beads". This has both positive and negative aspects. On one hand, although ENMs can predict the preferred directions for conformational change, the time and length scales of the motions (i.e., the magnitudes of the eigenvalues) are usually arbitrary. On the other, and paradoxically, this weakness reflects their major strength: ENMs are determined by protein shape, topology, and local packing density, and are thus insensitive to fine details. Despite these shortcomings and their dramatic simplicity, soft ENM modes are surprisingly accurate at predicting anharmonic, far-from-equilibrium transitions [20,40]. Together with the lack of a solvent and thus damping, this was initially a major point of controversy, questioning the validity of both NMA and its CG approximation [63]. What is the time and length scale of NMs? How can harmonic NMs capture anharmonic, damped and slow transitions over high energy barriers? It has been argued that proteins oscillate around the equilibrium, with energy increasing as they stretch along NMs' directions. This could elegantly agree with a dynamical systems perspective, as the Kolmogorov Arnold Moser (KAM) theorem assures the persistence of quasi-periodic motions under small perturbations [21,64]. Under this view, NMs would define major directions around a potential well, that hold relatively far from equilibrium. Following these, the high energy states reached would be further stretched and stabilized by different ligands or signals capable of "tipping" the free energy landscape (the so-called pre-existing equilibrium model [65,66], experimentally observed in enzymes [16]). Already in the 1990s, MD studies showed that indeed, the energy surface probed via simulations is well-approximated by a rescaled version of the harmonic potential [67,68]. Recent work has related anharmonicity to mode collectivity: low-frequency modes that are collective enough, remain harmonic even for large displacements and better correlate with experimental transitions [69]. The power of ENMs to explore the boundaries of free energy minima is thus being more and more recognized, to the point that they are now used to enhance sampling via MD [70]. Regarding the timescales question, it is clear that NMs cover all the protein motion timescales, from MHz (μs) large-scale motions to 1–10 THz (ps) backbone/atomic vibrations. However, the actual NM eigenvalues are typically meaningless and need rescaling, with few exceptions like the nearest-neighbors ED-ENM model [54]. Apart from this arbitrary amplitude of single modes, ENM–NMA tends to spread variance at higher frequencies in comparison to MD Essential Dynamics (ED) modes [58,71], probably as a consequence of the absence of damping. Our ED-ENM model [54], developed from database-wide comparisons with MD force fields, attempted to solve these issues by fitting spring functions not only to predict conformational changes but also to obtain realistic amplitudes for the eigenvalues and their distribution (i.e., the actual time and length scales in solution). This study also revealed that even extremely simple ENMs, just connecting the first three neighbors in the peptide chain, can predict MD and experimental flexibility, which critically depend on peptide backbone topology and local cohesiveness.

In brief, despite their many weaknesses—inconsistent parameterization, arbitrary time and length scales, lack of damping—the ability of ENMs to track functional large-scale motions—regardless of CG levels, spring definitions, or any sequence or local details—is stunning. Precisely in this fact lies the greatest physical insight they reveal: that proteins' overall packing, local connectivity, and shape determine intrinsic collective motions that poise them for function. These motions hold far beyond equilibrium and also across extremely long evolutionary scales, as we will discuss now.

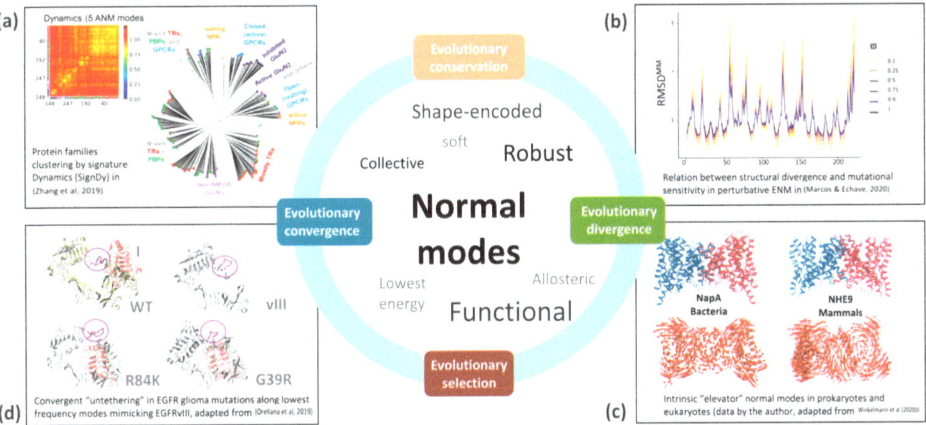

Figure 1. Shape-encoded ENM Normal Modes (NMs) and protein dynamics evolution examples from recent literature. (**a**) Signature Dynamics (SignDy) allows to build dynamics-based dendrograms comparable to those derived from sequence and structural similarity; see Ref. [72]. (**b**) Perturbative ENM suggests structural divergence relates more to mutational sensitivity (RMSD[MM]) than selection (ω), which only deepens the profiles. See details in Ref. [73]. (**c**) Prokaryotic–eukaryotic conservation of NMs coupled to function and (**d**) Mutational convergence to favor an NM transition towards an oncogenic intermediate characterized by the exposure of a cryptic epitope (purple circle). See also the discussion in Section 4 and further details in Figure 2 and Refs. [74,75], respectively. Images (**a,b**) have a Creative Commons Attribution License and (**c,d**) are adapted by the author from her work.

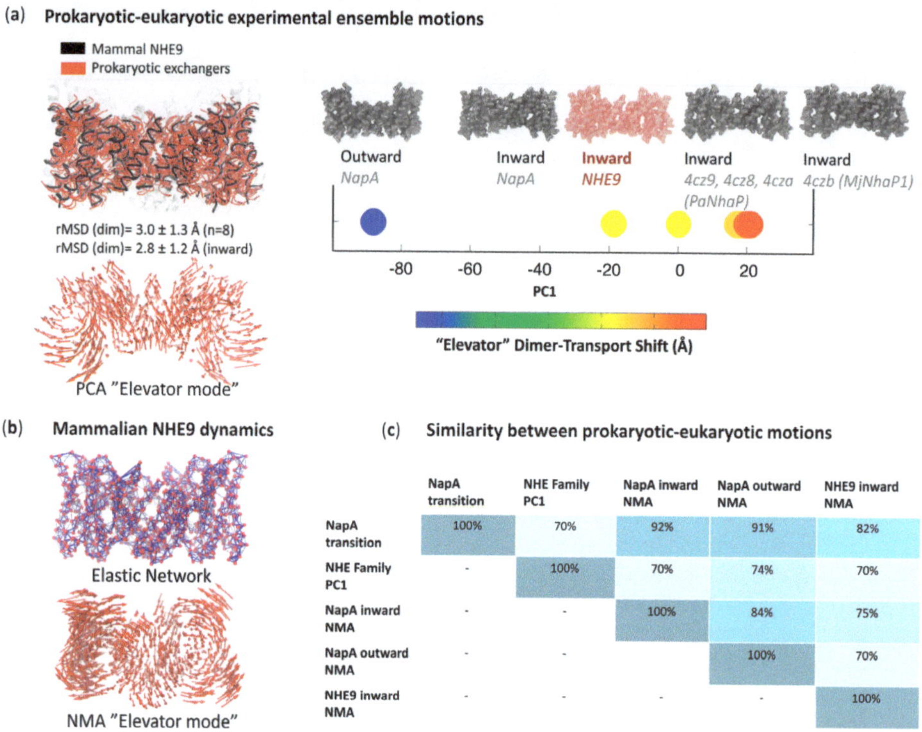

(a) Prokaryotic-eukaryotic experimental ensemble motions

(b) Mammalian NHE9 dynamics

(c) Similarity between prokaryotic-eukaryotic motions

	NapA transition	NHE Family PC1	NapA inward NMA	NapA outward NMA	NHE9 inward NMA
NapA transition	100%	70%	92%	91%	82%
NHE Family PC1	-	100%	70%	74%	70%
NapA inward NMA	-	-	100%	84%	75%
NapA outward NMA	-	-		100%	70%
NHE9 inward NMA	-	-	-	-	100%

Figure 2. A closer look at CPA exchangers' "elevator modes" conserved from bacteria to mammals. (a) **Left**: Core alignment between a mammalian exchanger, NHE9 (black) and distant bacterial homologs NapA, PanNhaP and MjNhaP1 (sequence identity ≈ 20%). The first principal component (PC1) of this ensemble of n = 8 structures renders the well-known "elevator-like" motion that distinguishes outward and inward states. **Right**: Projections onto PC1 of the experimental ensemble track the conformational inward-to-outward pathway and assigns the conformational status of the solved structures along it. (b) ENM of the mammalian NHE9 structure and derived "elevator-like" NM. (c) Similarity between NMs, PC1 and the prokaryotic NapA transition are all above 70%, despite the low sequence identities. Overlaps between vectorial spaces shaded in gradient; note that overlaps around 20% are considered random and from 40–50% significant. Adapted from figures and data by L. Orellana in Ref. [74], under the Creative Commons Attribution 4.0 License.

3. Lowest-Frequency Modes and Evolution

At the macroscopic level, we can easily appreciate how form, biological motion, and function evolve together under the laws of physics, shaping animal and plant morphologies [76]. Evolution seems to select the shapes best suited to perform functional motions. In the molecular world, if we assume the structure–motion–function paradigm, i.e., from motion comes function, it just follows to wonder whether evolution is selecting dynamics and resulting function rather than sequence or shape. Is there evidence of direct evolutionary pressure on protein motion? It is in this arena—where molecular evolution meets protein biophysics—that conformational dynamics becomes central [77]. Lowest-frequency modes allow for quantitative comparisons of the dynamics linked to function between similar cores [78], which are shedding new light on these questions.

Back in the 1980s, as soon as enough structures accumulated in the Protein Data Bank, it emerged that homologous proteins share similar folds, but this similarity wanes with increasing evolutionary distance [79,80]. Still, in practice, proteins with sequence similarities as low as 20% can display identical cores. The space of protein sequences is known to be much larger than that of structures, close to optimal [81] and restrained by the length, stability, and topology of each fold [82]. Importantly, from this fact, it also follows that structural folds, i.e., protein shapes, are highly robust against mutations. What about conformational spaces? ENMs have revealed that each structure preferentially samples a limited set of elemental motions; the shape determines the conformations/motions, and the motions define the function. Being defined by global shape, soft modes are also incredibly robust to perturbations like mutations [83] or local structural features, and therefore hold across protein families and even remote homologs. Hence, when two sequences have low but sizeable sequence similarity, they often share a common core, motions, and probably function [84]. Moreover, proteins sharing one similar conformation often share other conformations, i.e., their conformational spaces are conserved, a concept exploited to predict new conformers or model conformational changes [85]. Therefore, we could argue that, in the same way the sequence space is bigger than the structure space, the structure space is bigger than the motion space—and this inversely relates to fold and function robustness.

Based on mounting evidence from ENMs [20] and parallel studies on residue flexibility [86], protein global dynamics has been suggested to be maximally conserved versus sequence and structure. Nevertheless, the degree of conservation of conformational spaces as well as the contributing factors are unclear. Due to the entanglement of function, motion, and shape, together with protein biophysical and evolutionary constraints, the issue is intensely debated [87–89]. There are two central questions to be addressed: Is it function that primarily drives the conservation of dynamics? Or is it due to physical constraints such as stability, topology, local packing, etc., or properties like mode energies or robustness? What about evolutionary constraints such as population sizes, mutational rates or bias? In other words: are soft modes conserved because they are functional or because they are energetically "easy" and robust? Probably, the truth is in the middle.

Evidence for direct evolutionary pressure on normal modes is still scarce, as quantitative comparisons of functional dynamics are relatively recent [78]. It has been proposed that there is negative selection against the divergence of functionally important modes, while other studies suggest that they are conserved just because they are more robust to mutational perturbations (Figure 1a,b). Soon after ENMs were developed, it became evident that proteins with similar architecture shared similar motions [90]. Early studies on the evolution of soft modes, led by Ortiz and colleagues, focused on how structural cores modify their shape across homologous proteins [91–93]. These pioneering works revealed significant similarity in the conformational ensembles explored within a superfamily and the soft modes, i.e., proteins seem to evolutionarily diverge along soft modes or, vice versa, protein topology constrains evolutionary divergence. In parallel, Echave also showed that the lowest-frequency modes are conserved in homologous proteins [94], and there is a significant correlation between mode collectivity and its conservation [95]. The conservation of lowest-frequency modes is apparent in residue fluctuation patterns, which can be easily aligned for homologous proteins [96]. Some studies have also pointed out that protein sites evolve at different rates depending on properties such as their solvent accessibility, packing density, and flexibility [97,98]. In general, there is an inverse relation between local flexibility and evolutionary rates [99] i.e., exposed and flexible loops are less conserved than cores or rigid regions [100], which can act as hinges for global motions. Consequently, ENM analyses show clear correlations between sequence evolution and structural dynamics, especially relevant for hinge regions [100,101]. These rigid regions are so critical that hinge migration has been proposed as a mechanism for protein evolution [102]. Moreover, cancer and disease-related mutations tend to focus on hinge-like areas [103,104]. Therefore, ENM dynamics is a key predictor of functional impact for point mutations [105,106] as well as for insertions and deletions [107], further discussed below.

Importantly, even in the case of random mutations, structural changes correlate with the lowest frequency modes [108], as happens also for ensembles of the same protein determined in different experimental conditions [109]. Perturbative ENMs indicate that the conservation of soft modes might arise precisely from their robustness against mutations [110] and, conversely, structural divergence is proportional to mutational sensitivity [73]. Only mutations targeting critical regions such as rigid hinges could thus have the potential to change ENM mode patterns and function, causing either disease or driving evolution. The majority of changes would have no effect due to mode robustness, which would be the primary factor for evolutionary conservation. Apart from mode robustness, protein modularity and size also contribute to the overlap between the NMs and evolutionary modes and explain their low dimensionality, according to recent studies [111]. Altogether, these studies point out that biophysical properties are key for mode conservation.

Nevertheless, the functional motions observed experimentally seem to correlate with the soft modes more than expected based on just their amplitude and energies, indicating that selection plays a central role [112]. ENM studies indicate that selection guides sequence evolution to favor dynamical properties required for function, such as allosteric behavior or protein–protein interactions [113,114]. An exhaustive study by the Bahar group on nearly 27 K proteins representing 116 CATH superfamilies [72] characterized the cooperative mechanisms and convergent/divergent features that underlie the shared/differentiated dynamics of family members, developing an integrated pipeline to evaluate the signature dynamics of families based on ENMs (SignDy). They confirmed that global lowest-frequency modes of motion are conserved within a family, but there is a subset of motions that sharply distinguishes subfamilies at low-to-intermediate frequencies and is responsible for functional differentiation. Then, modulation of robust/conserved global dynamics via low-to-intermediate frequency fluctuations could be a versatile mechanism ensuring fold adaptability and subfamily specificity, subject to both positive and negative selection. Finally, taking one step further with this "selectionist" view, recent works have attempted to predict functional dynamics directly from sequence evolutionary couplings, skipping structures altogether [115].

4. Examples of Evolutionary Conservation, Convergence and Divergence

As we have seen, it is extremely difficult to disentangle the relevance of sequence, structure, and dynamics for evolutionary selection as they are intertwined. Database-wide comparative quantitative studies of protein dynamics are essential, but it is also important to keep in mind that, in the biological realm, "the devil can be in the details", and a closer look at key conserved systems can be illuminating to understand how and to what extent evolution polishes protein shape and motions (Figure 1c,d). This is especially true for proteins executing the most fundamental life processes, prevalent in almost all living species; it is also true for the disease almost intrinsic to the mechanisms of pluricellular life, cancer, which can be viewed as an evolutionary process in miniature [116]. For example, it is well known that cells critically depend on pH and ion homeostasis, as well as membrane transport. Unsurprisingly, solute carriers and ion channels mediating these processes are incredibly well conserved from bacteria to humans, despite diverging 2–4 billion years ago [117,118]. Despite very low sequence identities, prokaryotic and eukaryotic versions of proteins such as cation/proton antiporters (CPAs), major facilitator superfamily transporters (MFSs), or pentameric ligand-gated ion channels (PLGICs), are incredibly conserved from a structural and conformational point of view. CPAs mediate the exchange of protons and monovalent cations such as Na^+ or K^+, while MFS facilitates the movement of small solutes in response to gradients through cell membranes. Both MFSs and CPAs operate through an alternating-access mechanism, which requires a transition between states, where the substrate-binding site is exposed to opposite sides of the membrane alternately [119]. Structures show that MFSs follow a "rocker-switch" or "rocking bundle" mechanism, where the substrate-binding site is located at the interface of the so-called "transport" and "scaffold" domains. In contrast, CPAs work through an "elevator mechanism", where the substrate-binding site is confined largely to a single "transport" domain that traverses the membrane along a relatively rigid, immobile, and central "core". In the first, the barrier re-shapes and moves across the membrane while the substrate stays, while in the second, it stays at a fixed position, and it is the substrate that moves across it. Both transport mechanisms are dependent on large-scale transitions between the so-called "inward" and "outward" states. Remarkably, despite sequence identities around just 20%, structures of the mammal SLC/NHE CPA family of Na^+/H^+ exchangers bear striking similarity with prokaryotic ones, like those of bacterial *Thermus thermophilus* NapA, archaeal *Pyrococcus abyssii* PaNhaP or *Methanocaldococcus jannaschii* MjNhaP1. This makes it possible to extract a highly conserved structural core (756 residues per homodimer) to achieve an incredibly low RMSD near 3.0 ± 1.3 Å [74], which corresponds to the conformational transition tracked in the ensemble—when only one conformation is included, RMSD drops to 2 Å, close to thermal fluctuations (Figure 2 and Table 1). Both bacterial and mammal structures are thus solved in inward- and outward-facing states, and therefore, their core ensemble's main Principal Component (PC, see [26,60]) tracks the elevator motion responsible for transport. Significantly, this motion is also encoded in each one of the proteins: there is a high overlap (70–80%) between the transitions seen in the prokaryotic–eukaryotic ensemble and the lowest-frequency ENM modes from every individual member (Figure 2). Similarly, for MFSs, it is also possible to build a eukaryotic–prokaryotic "core" ensemble (353 residues) encompassing human, bovine, and rat GLUTs to *Plasmodium* PfHT1 or *Escherichia coli* XylE [120], that despite the sequence identity around 30% has an RMSD as low as 2.7 ± 1.2 Å and extremely similar rocking-bundle movements embedded on each structure. In the case of PLGICs, the notable resemblance between eukaryotic neurotransmitter channels and their simple prokaryotic counterparts like *Gloeobacter* GLIC has turned the latter into the perfect model to study gating mechanisms. As often happens with ancestral protein machines, their function (channel opening/closing) requires complex motions (extracellular blooming coupled to tilting/twisting of intracellular pore-gating helices), which are both embedded in their pentameric ring-like architecture and extremely conserved across evolution [55,121,122].

Table 1. Sequence, structural and dynamical similarity between mammalian NHE9 and bacterial proton exchanger NapA [1].

	Identity	Similarity	TM-Score
NHE9—NapA	22%	42%	0.82
Overlap NHE9—NapA NMA	75%		
Overlap NHE9—NapA X-ray transition	82%		

[1] Adapted from Ref. [74].

Finally, another example of evolutionary selection acting on conformation could be behind mutational asymmetries in cancer, which tend to target signaling proteins. Global dynamics is a predictor of missense mutation pathogenicity [105,123] and in cancer genes, it has been shown that mutations tend to cluster in specific functional spots and specifically hinge regions as determined via ENMs [104]. One striking example is the oncogene EGFR, which displays a puzzling tissue-specific mutational asymmetry. In brain glioblastoma (GBM), mutations are highly heterogenous but tend to cluster on the extracellular ligand-binding domain (ectodomain, ECD), even coexisting in the same tumor. In contrast, mutations in lung cancer concentrate in the intracellular kinase domain (KD), mostly focused on the catalytic cleft. This asymmetry results in intriguingly opposite responses to drugs binding to different KD conformers. Our ENM study of the ECD revealed that GBM mutations neatly cluster at hinge and interdomain regions, which control a large-scale conformational change of nearly 25 Å between the closed-unbound and open-bound states. Further MD simulations revealed that GBM mutations favor spontaneous ECD opening following the lowest frequency modes, to acquire a transient conformation known to exist but never trapped experimentally. This ENM/MD intermediate was validated through structural, in vitro, and in vivo experiments [75,124,125], is shared by missense mutants from different ECD hotspots, and mimics the configuration of the most frequent change in GBM, the deletion EGFRvIII (Figure 1d). Specifically, the first tandem repeat of EGFR is deleted in EGFRvIII but rotates in missense mutations. The ultimate goal of this remarkable structural "equivalence" or "convergence" trick is to allosterically activate the KD in a specific way, distinct from that favored by lung cancer mutations, which explains their different sensitivity to drugs. Importantly, lung and brain cancer mutations are known to activate different signaling pathways [126], and our ENM–MD studies suggest that this is directly governed by the different conformational dynamics they favor. On one hand, this could be an example of convergent evolution of missense mutations and deletions to achieve a similar functional outcome, driven by positive selection of those variants that explore the soft modes opening the structure in a "GBM-preferred" mode. On the other, the same protein, EGFR, apparently experiences divergent evolutionary trajectories in GBMs versus lung cancer to fine-tune its conformation and trigger cell growth in different niches—a potentially compelling case of evolution selecting lowest-frequency dynamics to modulate function.

In summary, the examples discussed above provide food for thought to question both the "selectionist-functional" view and the "biophysical-energetical" view of protein structure and dynamics evolution. Some works have focused on the interpretation of flexibility patterns under a predominantly evolutionary prism, while others favor the idea that the main cause of structural–dynamical divergence lies in the physical properties of proteins, such as their sensitivity to mutations. Observing the degree of conservation in ancestral proteins such as CPAs over scales of billions of years, despite having sequence identities in the "twilight" zone, strongly suggests a role for natural selection to keep key functional, structure-embedded mechanisms intact, especially for those proteins performing the most fundamental cellular tasks. These intrinsic motions have survived almost intact, from archaebacteria to the human species, probably because of both their biophysical robustness and their biological fitness. Conversely, the striking clustering of mutations observed in cancer proteins to modulate not only their intrinsic dynamics but also their interactions with other proteins, etc., shows that, at high mutational rates and under selection pressure, evolution can quickly remodel and adapt what we could call protein "molecular phenotypes" [77], directly determined by their conformational dynamics and the resulting biological function. Importantly, there is mounting evidence that even local dynamics coupled to processes such as enzyme catalysis show clear footprints of evolutionary selection [127–131]. Looking forward, there are wide opportunities to apply ENMs to deepen studies of molecular evolution, which can illuminate its connections with protein biophysics or even guide protein design [132]. From analysis of the conservation of flexible versus rigid regions and how they relate to function, to evolutionarily classifying proteins based on their shape-encoded dynamics rather than strict sequence information, ENMs will allow us to explore the interplay of flexibility and evolutionary changes in the different kingdoms to an extent never imagined before, even more thanks to the incredibly expanded structural spaces that AI has opened [133,134].

Overall, we foresee that as experimental and computational evidence accumulates, and the increasingly active research on ENMs and evolution develops, we might reach a new paradigm. One in which biomolecular dynamics and, specifically, the large-scale motions intrinsic to 3D structures, could effectively be considered what biologist Ernst Mayr called "an object of selection" [135] at the most basic, microscopic scale of life.

Funding: This research was funded by Karolinska Institute, the Swedish Foundations for Cancer Research (Cancerfonden Junior Investigator Award CF 21 0305 JIA and Project Grant CF 21 1471 Pj), the Swedish Scientific Research Council (Vetenskapsrådet, VR 2021-02248) and the Jeanssons, Hedlund and Sagen Foundations.

Data Availability Statement: Data used to generate the figures are available upon request.

Conflicts of Interest: The author declares no conflict of interest.

References

1. Anfinsen, C.B.; Haber, E.; Sela, M.; White, F.H. The Kinetics of Formation of Native Ribonuclease during Oxidation of the Reduced Polypeptide Chain. *Proc. Natl. Acad. Sci. USA* **1961**, *47*, 1309–1314. [CrossRef] [PubMed]
2. Daggett, V.; Fersht, A. The Present View of the Mechanism of Protein Folding. *Nat. Rev. Mol. Cell Biol.* **2003**, *4*, 497–502. [CrossRef] [PubMed]
3. Karplus, M.; McCammon, J.A. The Dynamics of Proteins. *Sci. Am.* **1986**, *254*, 42–51. [CrossRef] [PubMed]
4. Frank, J. New Opportunities Created by Single-Particle Cryo-EM: The Mapping of Conformational Space. *Biochemistry* **2018**, *57*, 888. [CrossRef] [PubMed]
5. James, L.C.; Tawfik, D.S. Conformational Diversity and Protein Evolution—A 60-Year-Old Hypothesis Revisited. *Trends Biochem. Sci.* **2003**, *28*, 361–368. [CrossRef] [PubMed]
6. Henzler-Wildman, K.; Kern, D. Dynamic Personalities of Proteins. *Nature* **2007**, *450*, 964–972. [CrossRef]
7. Karplus, M.; Kuriyan, J. Molecular Dynamics and Protein Function. *Proc. Natl. Acad. Sci. USA* **2005**, *102*, 6679–6685. [CrossRef]
8. Karplus, M.; McCammon, J.A. Molecular Dynamics Simulations of Biomolecules. *Nat. Struct. Biol.* **2002**, *9*, 646–652. [CrossRef]
9. Amaral, M.; Kokh, D.B.; Bomke, J.; Wegener, A.; Buchstaller, H.P.; Eggenweiler, H.M.; Matias, P.; Sirrenberg, C.; Wade, R.C.; Frech, M. Protein Conformational Flexibility Modulates Kinetics and Thermodynamics of Drug Binding. *Nat. Commun.* **2017**, *8*, 2276. [CrossRef]
10. Tuffery, P.; Derreumaux, P. Flexibility and Binding Affinity in Protein–Ligand, Protein–Protein and Multi-Component Protein Interactions: Limitations of Current Computational Approaches. *J. R. Soc. Interface* **2012**, *9*, 20–33. [CrossRef]
11. Teague, S.J. Implications of Protein Flexibility for Drug Discovery. *Nat. Rev. Drug Discov.* **2003**, *2*, 527–541. [CrossRef] [PubMed]
12. Changeux, J.-P.; Edelstein, S. Conformational Selection or Induced-Fit? 50 Years of Debate Resolved. *F1000 Biol. Rep.* **2011**, *3*, 1–15. [CrossRef] [PubMed]
13. Csermely, P.; Palotai, R.; Nussinov, R. Induced Fit, Conformational Selection and Independent Dynamic Segments: An Extended View of Binding Events. *Trends Biochem. Sci.* **2010**, *35*, 539–546. [CrossRef]
14. Thulasingam, M.; Orellana, L.; Nji, E.; Ahmad, S.; Rinaldo-Matthis, A.; Haeggström, J.Z. Crystal Structures of Human MGST2 Reveal Synchronized Conformational Changes Regulating Catalysis. *Nat. Commun.* **2021**, *12*, 5721. [CrossRef]
15. Mhashal, A.R.; Romero-Rivera, A.; Mydy, L.S.; Cristobal, J.R.; Gulick, A.M.; Richard, J.P.; Kamerlin, S.C.L. Modeling the Role of a Flexible Loop and Active Site Side Chains in Hydride Transfer Catalyzed by Glycerol-3-Phosphate Dehydrogenase. *ACS Catal.* **2020**, *10*, 11253–11267. [CrossRef]
16. Henzler-Wildman, K.A.; Thai, V.; Lei, M.; Ott, M.; Wolf-Watz, M.; Fenn, T.; Pozharski, E.; Wilson, M.A.; Petsko, G.A.; Karplus, M.; et al. Intrinsic Motions along an Enzymatic Reaction Trajectory. *Nature* **2007**, *450*, 838–844. [CrossRef] [PubMed]
17. Babu, M.M.; Van Der Lee, R.; De Groot, N.S.; Gsponer, J. Intrinsically Disordered Proteins: Regulation and Disease. *Curr. Opin. Struct. Biol.* **2011**, *21*, 432–440. [CrossRef]
18. Uversky, V.N. Intrinsically Disordered Proteins and Their "Mysterious" (Meta)Physics. *Front. Phys.* **2019**, *7*, 10. [CrossRef]
19. Elber, R.; Karplus, M. Multiple Conformational States of Proteins: A Molecular Dynamics Analysis of Myoglobin. *Science* **1987**, *235*, 318–321. [CrossRef]
20. Bahar, I.; Lezon, T.R.; Yang, L.-W.; Eyal, E. Global Dynamics of Proteins: Bridging between Structure and Function. *Annu. Rev. Biophys.* **2010**, *39*, 23–42. [CrossRef]
21. Bastolla, U. Computing Protein Dynamics from Protein Structure with Elastic Network Models. *Wiley Interdiscip. Rev. Comput. Mol. Sci.* **2014**, *4*, 488–503. [CrossRef]
22. López-Blanco, J.R.; Chacón, P. New Generation of Elastic Network Models. *Curr. Opin. Struct. Biol.* **2016**, *37*, 46–53. [CrossRef] [PubMed]
23. Sanejouand, Y.-H. Elastic Network Models: Theoretical and Empirical Foundations. *Network* **2011**, *26*, 601–616.
24. Bonomi, M.; Vendruscolo, M. Determination of Protein Structural Ensembles Using Cryo-Electron Microscopy. *Curr. Opin. Struct. Biol.* **2019**, *56*, 37–450. [CrossRef] [PubMed]
25. Krieger, J.M.; Sorzano, C.O.S.; Carazo, J.M.; Bahar, I. Protein Dynamics Developments for the Large Scale and CryoEM: Case Study of ProDy 2.0. *Acta Cryst. D Struct. Biol.* **2022**, *78*, 399–409. [CrossRef]
26. Orellana, L. Large-Scale Conformational Changes and Protein Function: Breaking the in Silico Barrier. *Front. Mol. Biosci.* **2019**, *6*, 117. [CrossRef]
27. Herzberg, G. *Molecular Spectra and Molecular Structure*; D. Van Nostrand Company, Inc.: Princeton, NJ, USA, 1945.
28. Wilson, E.B.; Decius, J.C.; Cross, P.C. *Molecular Vibrations: The Theory of Infrared and Raman Vibrational Spectra*; McGraw-Hill: New York, NY, USA, 1955.
29. Brooks, B. Harmonic Dynamics of Proteins: Normal Modes and Fluctuations in Bovine Pancreatic Trypsin Inhibitor. *Proc. Natl. Acad. Sci. USA* **1983**, *80*, 6571–6575. [CrossRef]
30. Go, N.; Noguti, T.; Nishikawa, T. Dynamics of a Small Globular Protein in Terms of Low-Frequency Vibrational Modes. *Proc. Natl. Acad. Sci. USA* **1983**, *80*, 3696–3700. [CrossRef]
31. Levitt, M.; Sander, C.; Stern, P.S. The normal modes of a protein: Native bovine pancreatic trypsin inhibitor. *Int. J. Quantum Chem.* **1983**, *24*, 181–199. [CrossRef]
32. Noguti, T.; Gō, N. Collective Variable Description of Small-Amplitude Conformational Fluctuations in a Globular Protein. *Nature* **1982**, *296*, 776–778. [CrossRef]

33. Tasumi, M.; Takeuchi, H.; Ataka, S.; Dwivedi, A.M.; Krimm, S. Normal Vibrations of Proteins: Glucagon. *Biopolymers* **1982**, *21*, 711–714. [CrossRef] [PubMed]
34. Orozco, M.; Orellana, L.; Hospital, A.; Naganathan, A.N.; Emperador, A.; Carrillo, O.; Gelpí, J.L. Coarse-Grained Representation of Protein Flexibility. Foundations, Successes, and Shortcomings. *Adv. Protein Chem. Struct. Biol.* **2011**, *85*, 183–215. [CrossRef] [PubMed]
35. Flory, P.J.; Gordon, M.; McCrum, N.G. Statistical Thermodynamics of Random Networks [and Discussion]. *Proc. R. Soc. A Math. Phys. Eng. Sci.* **1976**, *351*, 351–380. [CrossRef]
36. Rouse, P.E. A Theory of the Linear Viscoelastic Properties of Dilute Solutions of Coiling Polymers. *J. Chem. Phys.* **1953**, *21*, 1272. [CrossRef]
37. Tirion, M. Large Amplitude Elastic Motions in Proteins from a Single-Parameter, Atomic Analysis. *Phys. Rev. Lett.* **1996**, *77*, 1905–1908. [CrossRef]
38. Bahar, I.; Atilgan, A.R.; Erman, B. Direct Evaluation of Thermal Fluctuations in Proteins Using a Single-Parameter Harmonic Potential. *Fold. Des.* **1997**, *2*, 173–181. [CrossRef]
39. Atilgan, A.R.; Durell, S.R.; Jernigan, R.L.; Demirel, M.C.; Keskin, O.; Bahar, I. Anisotropy of Fluctuation Dynamics of Proteins with an Elastic Network Model. *Biophys. J.* **2001**, *80*, 505–515. [CrossRef]
40. Bauer, J.A.; Pavlovíc, J.; Bauerová-Hlinková, V. Normal Mode Analysis as a Routine Part of a Structural Investigation. *Molecules* **2019**, *24*, 3293. [CrossRef]
41. Dobbins, S.E.; Lesk, V.I.; Sternberg, M.J.E. Insights into Protein Flexibility: The Relationship between Normal Modes and Conformational Change upon Protein-Protein Docking. *Proc. Natl. Acad. Sci. USA* **2008**, *105*, 10390–10395. [CrossRef]
42. Petrone, P.; Pande, V.S. Can Conformational Change Be Described by Only a Few Normal Modes? *Biophys. J.* **2006**, *90*, 1583–1593. [CrossRef]
43. Stein, A.; Rueda, M.; Panjkovich, A.; Orozco, M.; Aloy, P. A Systematic Study of the Energetics Involved in Structural Changes upon Association and Connectivity in Protein Interaction Networks. *Structure* **2011**, *19*, 881–889. [CrossRef] [PubMed]
44. Yang, L.; Song, G.; Jernigan, R.L. How Well Can We Understand Large-Scale Protein Motions Using Normal Modes of Elastic Network Models? *Biophys. J.* **2007**, *93*, 920–929. [CrossRef] [PubMed]
45. Vu, H.T.; Zhang, Z.; Tehver, R.; Thirumalai, D. Plus and Minus Ends of Microtubules Respond Asymmetrically to Kinesin Binding by a Long-Range Directionally Driven Allosteric Mechanism. *Sci. Adv.* **2022**, *8*, eabn0856. [CrossRef] [PubMed]
46. Kolan, D.; Fonar, G.; Samson, A.O. Elastic Network Normal Mode Dynamics Reveal the GPCR Activation Mechanism. *Proteins Struct. Funct. Bioinform.* **2014**, *82*, 579–586. [CrossRef]
47. Bahar, I. On the Functional Significance of Soft Modes Predicted by Coarse-Grained Models for Membrane Proteins. *J. Gen. Physiol.* **2010**, *135*, 563–573. [CrossRef]
48. Isin, B.; Rader, A.J.; Dhiman, H.K.; Klein-Seetharaman, J.; Bahar, I. Predisposition of the Dark State of Rhodopsin to Functional Changes in Structure. *Proteins Struct. Funct. Bioinform.* **2006**, *65*, 970–983. [CrossRef]
49. Gerstein, M.; Krebs, W. A Database of Macromolecular Motions. *Nucleic Acids Res.* **1998**, *26*, 4280–4290. [CrossRef]
50. Krebs, W.G.; Alexandrov, V.; Wilson, C.A.; Echols, N.; Yu, H.; Gerstein, M. Normal Mode Analysis of Macromolecular Motions in a Database Framework: Developing Mode Concentration as a Useful Classifying Statistic. *Proteins* **2002**, *48*, 682–695. [CrossRef]
51. Tama, F.; Sanejouand, Y.H. Conformational Change of Proteins Arising from Normal Mode Calculations. *Protein Eng.* **2001**, *14*, 1–6. [CrossRef]
52. Alexandrov, V. Normal Modes for Predicting Protein Motions: A Comprehensive Database Assessment and Associated Web Tool. *Protein Sci.* **2005**, *14*, 633–643. [CrossRef]
53. Gur, M.; Zomot, E.; Bahar, I. Global Motions Exhibited by Proteins in Micro- to Milliseconds Simulations Concur with Anisotropic Network Model Predictions. *J. Chem. Phys.* **2013**, *139*, 121912. [CrossRef]
54. Orellana, L.; Rueda, M.; Ferrer-Costa, C.; Lopez-Blanco, J.R.; Chacón, P.; Orozco, M. Approaching Elastic Network Models to Molecular Dynamics Flexibility. *J. Chem. Theory Comput.* **2010**, *6*, 2910–2923. [CrossRef] [PubMed]
55. Orellana, L.; Yoluk, O.; Carrillo, O.; Orozco, M.; Lindahl, E. Prediction and Validation of Protein Intermediate States from Structurally Rich Ensembles and Coarse-Grained Simulations. *Nat. Commun.* **2016**, *7*, 12575. [CrossRef] [PubMed]
56. Yang, L.; Song, G.; Carriquiry, A.; Jernigan, R.L. Close Correspondence between the Motions from Principal Component Analysis of Multiple HIV-1 Protease Structures and Elastic Network Modes. *Structure* **2008**, *16*, 321–330. [CrossRef] [PubMed]
57. Rueda, M.; Chacón, P.; Orozco, M. Thorough Validation of Protein Normal Mode Analysis: A Comparative Study with Essential Dynamics. *Structure* **2007**, *15*, 565–575. [CrossRef]
58. Daidone, I.; Amadei, A. Essential Dynamics: Foundation and Applications. *Wiley Interdiscip. Rev. Comput. Mol. Sci.* **2012**, *2*, 762–770. [CrossRef]
59. Jollife, I.T.; Cadima, J. Principal Component Analysis: A Review and Recent Developments. *Philos. Trans. R. Soc. A Math. Phys. Eng. Sci.* **2016**, *374*, 20150202. [CrossRef]
60. Kitao, A. Principal Component Analysis and Related Methods for Investigating the Dynamics of Biological Macromolecules. *J* **2022**, *5*, 298–317. [CrossRef]
61. Rueda, M.; Ferrer-Costa, C.; Meyer, T.; Pérez, A.; Camps, J.; Hospital, A.; Gelpí, J.L.; Orozco, M. A Consensus View of Protein Dynamics. *Proc. Natl. Acad. Sci. USA* **2007**, *104*, 796–801. [CrossRef]

62. Meyer, T.; D'Abramo, M.; Hospital, A.; Rueda, M.; Ferrer-Costa, C.; Pérez, A.; Carrillo, O.; Camps, J.; Fenollosa, C.; Repchevsky, D.; et al. MoDEL (Molecular Dynamics Extended Library): A Database of Atomistic Molecular Dynamics Trajectories. *Structure* **2010**, *18*, 1399–1409. [CrossRef]

63. Ma, J. Usefulness and Limitations of Normal Mode Analysis in Modeling Dynamics of Biomolecular Complexes. *Structure* **2005**, *13*, 373–380. [CrossRef] [PubMed]

64. Hubbard, J.H. The KAM Theorem. In *Kolmogorov's Heritage in Mathematics*; Charpentier, É., Lesne, A., Nikolski, N.K., Eds.; Springer: Berlin/Heidelberg, Germany, 2007; pp. 215–238. ISBN 978-3-540-36351-4.

65. Kern, D.; Zuiderweg, E.R. The Role of Dynamics in Allosteric Regulation. *Curr. Opin. Struct. Biol.* **2003**, *13*, 748–757. [CrossRef] [PubMed]

66. Goh, C.-S.; Milburn, D.; Gerstein, M. Conformational Changes Associated with Protein-Protein Interactions. *Curr. Opin. Struct. Biol.* **2004**, *14*, 104–109. [CrossRef]

67. Hayward, S.; Kitao, A.; Go, N. Harmonic and Anharmonic Aspects in the Dynamics of BPTI: A Normal Mode Analysis and Principal Component Analysis. *Protein Sci. A Publ. Protein Soc.* **1994**, *3*, 936–943. [CrossRef] [PubMed]

68. Hayward, S.; Kitao, A.; Go, N. Harmonicity and Anharmonicity in Protein Dynamics: A Normal Mode Analysis and Principal Component Analysis. *Proteins* **1995**, *23*, 177–186. [CrossRef]

69. Dehouck, Y.; Bastolla, U. Why Are Large Conformational Changes Well Described by Harmonic Normal Modes? *Biophys. J.* **2021**, *120*, 5343–5354. [CrossRef]

70. Kaynak, B.T.; Krieger, J.M.; Dudas, B.; Dahmani, Z.L.; Costa, M.G.S.; Balog, E.; Scott, A.L.; Doruker, P.; Perahia, D.; Bahar, I. Sampling of Protein Conformational Space Using Hybrid Simulations: A Critical Assessment of Recent Methods. *Front. Mol. Biosci.* **2022**, *9*, 832847. [CrossRef]

71. Amadei, A.; Linssen, A.B.; Berendsen, H.J. Essential Dynamics of Proteins. *Proteins* **1993**, *17*, 412–425. [CrossRef] [PubMed]

72. Zhang, S.; Li, H.; Krieger, J.M.; Bahar, I. Shared Signature Dynamics Tempered by Local Fluctuations Enables Fold Adaptability and Specificity. *Mol. Biol. Evol.* **2019**, *36*, 2053–2068. [CrossRef]

73. Marcos, M.L.; Echave, J. The Variation among Sites of Protein Structure Divergence Is Shaped by Mutation and Scaled by Selection. *Curr. Res. Struct. Biol.* **2020**, *2*, 156–163. [CrossRef]

74. Winkelmann, I.; Matsuoka, R.; Meier, P.F.; Shutin, D.; Zhang, C.; Orellana, L.; Sexton, R.; Landreh, M.; Robinson, C.V.; Beckstein, O.; et al. Structure and Elevator Mechanism of the Mammalian Sodium/Proton Exchanger NHE9. *EMBO J.* **2020**, *39*, 4541–4559. [CrossRef] [PubMed]

75. Orellana, L.; Thorne, A.H.; Lema, R.; Gustavsson, J.; Parisian, A.D.; Hospital, A.; Cordeiro, T.N.; Bernadó, P.; Scott, A.M.; Brun-Heath, I.; et al. Oncogenic Mutations at the EGFR Ectodomain Structurally Converge to Remove a Steric Hindrance on a Kinase-Coupled Cryptic Epitope. *Proc. Natl. Acad. Sci. USA* **2019**, *116*, 10009–10018. [CrossRef]

76. Muñoz, M.M.; Price, S.A. The Future Is Bright for Evolutionary Morphology and Biomechanics in the Era of Big Data. *Integr. Comp. Biol.* **2019**, *59*, 599–603. [CrossRef] [PubMed]

77. Sikosek, T.; Chan, H.S. Biophysics of Protein Evolution and Evolutionary Protein Biophysics. *J. R. Soc. Interface* **2014**, *11*, 20140419. [CrossRef] [PubMed]

78. Fuglebakk, E.; Tiwari, S.P.; Reuter, N. Comparing the Intrinsic Dynamics of Multiple Protein Structures Using Elastic Network Models. *Biochim. Biophys. Acta—Gen. Subj.* **2015**, *1850*, 911–922. [CrossRef] [PubMed]

79. Bordin, N.; Sillitoe, I.; Lees, J.G.; Orengo, C. Tracing Evolution Through Protein Structures: Nature Captured in a Few Thousand Folds. *Front. Mol. Biosci.* **2021**, *8*, 668184. [CrossRef]

80. Chothia, C.; Lesk, A.M. The Relation between the Divergence of Sequence and Structure in Proteins. *EMBO J.* **1986**, *5*, 823–826. [CrossRef]

81. Kuhlman, B.; Baker, D. Native Protein Sequences Are Close to Optimal for Their Structures. *Proc. Natl. Acad. Sci. USA* **2000**, *97*, 10383–10388. [CrossRef]

82. Koehl, P.; Levitt, M. Protein Topology and Stability Define the Space of Allowed Sequences. *Proc. Natl. Acad. Sci. USA* **2002**, *99*, 1280–1285. [CrossRef]

83. Zheng, W.; Brooks, B.R.; Thirumalai, D. Low-Frequency Normal Modes That Describe Allosteric Transitions in Biological Nanomachines Are Robust to Sequence Variations. *Proc. Natl. Acad. Sci. USA* **2006**, *103*, 7664–7669. [CrossRef]

84. Hensen, U.; Meyer, T.; Haas, J.; Rex, R.; Vriend, G.; Grubmüller, H. Exploring Protein Dynamics Space: The Dynasome as the Missing Link between Protein Structure and Function. *PLoS ONE* **2012**, *7*, e33931. [CrossRef] [PubMed]

85. Narunsky, A.; Nepomnyachiy, S.; Ashkenazy, H.; Kolodny, R.; Ben-Tal, N. ConTemplate Suggests Possible Alternative Conformations for a Query Protein of Known Structure. *Structure* **2015**, *23*, 2162–2170. [CrossRef] [PubMed]

86. Zsolyomi, F.; Ambrus, V.; Fuxreiter, M. Patterns of Dynamics Comprise a Conserved Evolutionary Trait. *J. Mol. Biol.* **2020**, *432*, 497–507. [CrossRef] [PubMed]

87. Bastolla, U.; Dehouck, Y.; Echave, J. What Evolution Tells Us about Protein Physics, and Protein Physics Tells Us about Evolution. *Curr. Opin. Struct. Biol.* **2017**, *42*, 59–66. [CrossRef] [PubMed]

88. Liberles, D.A.; Teichmann, S.A.; Bahar, I.; Bastolla, U.; Bloom, J.; Bornberg-Bauer, E.; Colwell, L.J.; De Koning, A.P.J.; Dokholyan, N.V.; Echave, J.; et al. The Interface of Protein Structure, Protein Biophysics, and Molecular Evolution. *Protein Sci.* **2012**, *21*, 769–785. [CrossRef] [PubMed]

89. Tiwari, S.P.; Reuter, N. Conservation of Intrinsic Dynamics in Proteins—What Have Computational Models Taught Us? *Curr. Opin. Struct. Biol.* **2018**, *50*, 75–81. [CrossRef] [PubMed]
90. Keskin, O.; Jernigan, R.L.; Bahar, I. Proteins with Similar Architecture Exhibit Similar Large-Scale Dynamic Behavior. *Biophys. J.* **2000**, *78*, 2093–2106. [CrossRef]
91. Leo-Macias, A.; Lopez-Romero, P.; Lupyan, D.; Zerbino, D.; Ortiz, A.R. An Analysis of Core Deformations in Protein Superfamilies. *Biophys. J.* **2005**, *88*, 1291–1299. [CrossRef]
92. Velázquez-Muriel, J.A.; Rueda, M.; Cuesta, I.; Pascual-Montano, A.; Orozco, M.; Carazo, J.-M. Comparison of Molecular Dynamics and Superfamily Spaces of Protein Domain Deformation. *BMC Struct. Biol.* **2009**, *9*, 6. [CrossRef]
93. Leo-Macias, A.; Lopez-Romero, P.; Lupyan, D.; Zerbino, D.; Ortiz, A.R. Core Deformations in Protein Families: A Physical Perspective. *Biophys. Chem.* **2005**, *115*, 125–128. [CrossRef]
94. Maguid, S.; Fernandez-Alberti, S.; Ferrelli, L.; Echave, J. Exploring the Common Dynamics of Homologous Proteins. Application to the Globin Family. *Biophys. J.* **2005**, *89*, 3–13. [CrossRef] [PubMed]
95. Maguid, S.; Fernández-Alberti, S.; Parisi, G.; Echave, J. Evolutionary Conservation of Protein Backbone Flexibility. *J. Mol. Evol.* **2006**, *63*, 448–457. [CrossRef] [PubMed]
96. Skjaerven, L.; Yao, X.Q.; Scarabelli, G.; Grant, B.J. Integrating Protein Structural Dynamics and Evolutionary Analysis with Bio3D. *BMC Bioinform.* **2014**, *15*, 399. [CrossRef]
97. Franzosa, E.A.; Xia, Y. Structural Determinants of Protein Evolution Are Context-Sensitive at the Residue Level. *Mol. Biol. Evol.* **2009**, *26*, 2387–2395. [CrossRef] [PubMed]
98. Huang, T.-T.; del Valle Marcos, M.L.; Hwang, J.-K.; Echave, J. A Mechanistic Stress Model of Protein Evolution Accounts for Site-Specific Evolutionary Rates and Their Relationship with Packing Density and Flexibility. *BMC Evol. Biol.* **2014**, *14*, 78. [CrossRef]
99. Marsh, J.A.; Teichmann, S.A. Parallel Dynamics and Evolution: Protein Conformational Fluctuations and Assembly Reflect Evolutionary Changes in Sequence and Structure. *BioEssays* **2014**, *36*, 209–218. [CrossRef]
100. Dong, Z.; Zhou, H.; Tao, P. Combining Protein Sequence, Structure, and Dynamics: A Novel Approach for Functional Evolution Analysis of PAS Domain Superfamily. *Protein Sci.* **2018**, *27*, 421–430. [CrossRef]
101. Liu, Y.; Bahar, I. Sequence Evolution Correlates with Structural Dynamics. *Mol. Biol. Evol.* **2012**, *29*, 2253–2263. [CrossRef]
102. Campitelli, P.; Modi, T.; Kumar, S.; Ozkan, S.B. The Role of Conformational Dynamics and Allostery in Modulating Protein Evolution. *Annu. Rev. Biophys.* **2020**, *49*, 267–288. [CrossRef]
103. Nevin Gerek, Z.; Kumar, S.; Banu Ozkan, S. Structural Dynamics Flexibility Informs Function and Evolution at a Proteome Scale. *Evol. Appl.* **2013**, *6*, 423–433. [CrossRef]
104. Sayılgan, J.F.; Haliloğlu, T.; Gönen, M. Protein Dynamics Analysis Reveals That Missense Mutations in Cancer-Related Genes Appear Frequently on Hinge-Neighboring Residues. *Proteins Struct. Funct. Bioinform.* **2019**, *87*, 512–519. [CrossRef] [PubMed]
105. Ponzoni, L.; Bahar, I. Structural Dynamics Is a Determinant of the Functional Significance of Missense Variants. *Proc. Natl. Acad. Sci. USA* **2018**, *115*, 4164–4169. [CrossRef]
106. Frappier, V.; Najmanovich, R.J. A Coarse-Grained Elastic Network Atom Contact Model and Its Use in the Simulation of Protein Dynamics and the Prediction of the Effect of Mutations. *PLoS Comput. Biol.* **2014**, *10*, e1003569. [CrossRef]
107. Banerjee, A.; Bahar, I. Structural Dynamics Predominantly Determine the Adaptability of Proteins to Amino Acid Deletions. *Int. J. Mol. Sci.* **2023**, *24*, 8450. [CrossRef]
108. Echave, J. Evolutionary Divergence of Protein Structure: The Linearly Forced Elastic Network Model. *Chem. Phys. Lett.* **2008**, *457*, 413–416. [CrossRef]
109. Echave, J.; Fernández, F.M. A Perturbative View of Protein Structural Variation. *Proteins Struct. Funct. Bioinform.* **2010**, *78*, 173–180. [CrossRef]
110. Echave, J. Why Are the Low-Energy Protein Normal Modes Evolutionarily Conserved? *Pure Appl. Chem.* **2012**, *84*, 1931–1937. [CrossRef]
111. Tang, Q.-Y.; Kaneko, K. Dynamics-Evolution Correspondence in Protein Structures. *Phys. Rev. Lett.* **2021**, *127*, 098103. [CrossRef]
112. Dos Santos, H.G.; Klett, J.; Méndez, R.; Bastolla, U. Characterizing Conformation Changes in Proteins through the Torsional Elastic Response. *Biochim. Biophys. Acta* **2013**, *1834*, 836–846. [CrossRef]
113. Haliloglu, T.; Bahar, I. Adaptability of Protein Structures to Enable Functional Interactions and Evolutionary Implications. *Curr. Opin. Struct. Biol.* **2015**, *35*, 17–23. [CrossRef] [PubMed]
114. Zhang, Y.; Doruker, P.; Kaynak, B.; Zhang, S.; Krieger, J.; Li, H.; Bahar, I. Intrinsic Dynamics Is Evolutionarily Optimized to Enable Allosteric Behavior. *Curr. Opin. Struct. Biol.* **2020**, *62*, 14–21. [CrossRef] [PubMed]
115. Jia, K.; Kilinc, M.; Jernigan, R.L. Functional Protein Dynamics Directly from Sequences. *J. Phys. Chem. B* **2023**, *127*, 1914–1921. [CrossRef] [PubMed]
116. Nowell, P.C. The Clonal Evolution of Tumor Cell Populations. *Science* **1976**, *194*, 23–28. [CrossRef]
117. Hedges, S.B.; Chen, H.; Kumar, S.; Wang, D.Y.; Thompson, A.S.; Watanabe, H. A Genomic Timescale for the Origin of Eukaryotes. *BMC Evol. Biol.* **2001**, *1*, 4. [CrossRef] [PubMed]
118. Long, X.; Xue, H.; Wong, J.T.-F. Descent of Bacteria and Eukarya From an Archaeal Root of Life. *Evol. Bioinform. Online* **2020**, *16*, 1176934320908267. [CrossRef]

119. Drew, D.; Boudker, O. Shared Molecular Mechanisms of Membrane Transporters. *Annu. Rev. Biochem.* **2016**, *85*, 543–572. [CrossRef] [PubMed]
120. Qureshi, A.A.; Suades, A.; Matsuoka, R.; Brock, J.; McComas, S.E.; Nji, E.; Orellana, L.; Claesson, M.; Delemotte, L.; Drew, D. The Molecular Basis for Sugar Import in Malaria Parasites. *Nature* **2020**, *578*, 321–325. [CrossRef]
121. Howard, R.J. Elephants in the Dark: Insights and Incongruities in Pentameric Ligand-Gated Ion Channel Models. *J. Mol. Biol.* **2021**, *433*, 167128. [CrossRef]
122. Mhashal, A.R.; Yoluk, O.; Orellana, L. Exploring the Conformational Impact of Novel Glycine Receptor Mutations through Coarse-Grained Analysis and Atomistic Simulations. *Front. Mol. Biosci.* **2022**, *9*, 890851. [CrossRef]
123. Ponzoni, L.; Peñaherrera, D.A.; Oltvai, Z.N.; Bahar, I. Rhapsody: Predicting the Pathogenicity of Human Missense Variants. *Bioinformatics* **2020**, *36*, 3084–3092. [CrossRef]
124. Orellana, L. Convergence of EGFR Glioblastoma Mutations: Evolution and Allostery Rationalizing Targeted Therapy. *Mol. Cell. Oncol.* **2019**, *6*, e1630798. [CrossRef] [PubMed]
125. Orellana, L.; Hospital, A.; Orozco, M. Oncogenic Mutations of the EGF-Receptor Ectodomain Reveal an Unexpected Mechanism for Ligand-Independent Activation. *bioRxiv* **2014**. [CrossRef]
126. Uribe, M.L.; Marrocco, I.; Yarden, Y. EGFR in Cancer: Signaling Mechanisms, Drugs, and Acquired Resistance. *Cancers* **2021**, *13*, 2748. [CrossRef] [PubMed]
127. Lai, J.; Jin, J.; Kubelka, J.; Liberles, D.A. A Phylogenetic Analysis of Normal Modes Evolution in Enzymes and Its Relationship to Enzyme Function. *J. Mol. Biol.* **2012**, *422*, 442–459. [CrossRef]
128. Petrovic, D.; Risso, V.A.; Kamerlin, S.C.L.; Sanchez-Ruiz, J.M. Conformational Dynamics and Enzyme Evolution. *J. R. Soc. Interface* **2018**, *15*, 20180330. [CrossRef]
129. Narayanan, C.; Bernard, D.N.; Bafna, K.; Gagné, D.; Chennubhotla, C.S.; Doucet, N.; Agarwal, P.K. Conservation of Dynamics Associated with Biological Function in an Enzyme Superfamily. *Structure* **2018**, *26*, 426–436.e3. [CrossRef]
130. Ramanathan, A.; Agarwal, P.K. Evolutionarily Conserved Linkage between Enzyme Fold, Flexibility, and Catalysis. *PLoS Biol.* **2011**, *9*, e1001193. [CrossRef]
131. Carnevale, V.; Raugei, S.; Micheletti, C.; Carloni, P. Convergent Dynamics in the Protease Enzymatic Superfamily. *J. Am. Chem. Soc.* **2006**, *128*, 9766–9772. [CrossRef]
132. Campbell, E.C.; Correy, G.J.; Mabbitt, P.D.; Buckle, A.M.; Tokuriki, N.; Jackson, C.J. Laboratory Evolution of Protein Conformational Dynamics. *Curr. Opin. Struct. Biol.* **2018**, *50*, 49–57. [CrossRef]
133. Jumper, J.; Evans, R.; Pritzel, A.; Green, T.; Figurnov, M.; Ronneberger, O.; Tunyasuvunakool, K.; Bates, R.; Žídek, A.; Potapenko, A.; et al. Highly Accurate Protein Structure Prediction with AlphaFold. *Nature* **2021**, *596*, 583–589. [CrossRef]
134. Lin, Z.; Akin, H.; Rao, R.; Hie, B.; Zhu, Z.; Lu, W.; Smetanin, N.; Verkuil, R.; Kabeli, O.; Shmueli, Y.; et al. Evolutionary-Scale Prediction of Atomic-Level Protein Structure with a Language Model. *Science* **2023**, *379*, 1123–1130. [CrossRef] [PubMed]
135. Mayr, E. The Objects of Selection. *Proc. Natl. Acad. Sci. USA* **1997**, *94*, 2091–2094. [CrossRef] [PubMed]

MDPI AG
Grosspeteranlage 5
4052 Basel
Switzerland
Tel.: +41 61 683 77 34

Applied Sciences Editorial Office
E-mail: applsci@mdpi.com
www.mdpi.com/journal/applsci